Light–Material Interactions and Applications in Optoelectronic Devices

Online at: https://doi.org/10.1088/978-0-7503-6099-9

IOP Series in Advances in Optics, Photonics and Optoelectronics

SERIES EDITOR

Professor Rajpal S Sirohi Consultant Scientist

About the Editor

Rajpal S Sirohi is currently working as a faculty member in the Department of Physics, Alabama A&M University, Huntsville, AL, USA. Prior to this, he was a consultant scientist at the Indian Institute of Science, Bangalore, and before that he was Chair Professor in the Department of Physics, Tezpur University, Assam. During 2000–2011, he was an academic administrator, being vice-chancellor to a couple of universities and the director of the Indian Institute of Technology, Delhi. He is the recipient of many international and national awards and the author of more than 400 papers. Dr Sirohi is involved with research concerning optical metrology, optical instrumentation, holography, and the speckle phenomena.

About the series

Optics, photonics, and optoelectronics are enabling technologies in many branches of science, engineering, medicine, and agriculture. These technologies have reshaped our outlook and our ways of interacting with each other, and have brought people closer together. They help us to understand many phenomena better and provide deeper insight into the functioning of nature. Further, these technologies themselves are evolving at a rapid rate. Their applications encompass very large spatial scales, from nanometers to the astronomical scale, and a very large temporal range, from picoseconds to billions of years. This series on advances in optics, photonics, and optoelectronics aims to cover topics that are of interest to both academia and industry. Some of the topics to be covered by the books in this series include biophotonics and medical imaging, devices, electromagnetics, fiber optics, information storage, instrumentation, light sources, charge-coupled devices (CCDs) and complementary metal oxide semiconductor (CMOS) imagers, metamaterials, optical metrology, optical networks, photovoltaics, free-form optics and its evaluation, singular optics, cryptography, and sensors.

About IOP ebooks

The authors are encouraged to take advantage of the features made possible by electronic publication to enhance the reader experience through the use of color, animation, and video and by incorporating supplementary files in their work.

A full list of titles published in this series can be found here: https://iopscience.iop.org/bookListInfo/series-on-advances-in-optics-photonics-and-optoelectronics.

Light–Material Interactions and Applications in Optoelectronic Devices

Anjali Chandel
Sheng Hsiung Chang
Department of Optoelectronics and Materials Technology,
National Taiwan Ocean University, Keelung, Taiwan

IOP Publishing, Bristol, UK

Anjali Chandel and Sheng Hsiung Chang have asserted their right to be identified as the authors of this work in accordance with sections 77 and 78 of the Copyright, Designs and Patents Act 1988.

ISBN   978-0-7503-6099-9 (ebook)
ISBN   978-0-7503-6097-5 (print)
ISBN   978-0-7503-6100-2 (myPrint)
ISBN   978-0-7503-6098-2 (mobi)

DOI   10.1088/978-0-7503-6099-9

Version: 20241201

IOP ebooks

British Library Cataloguing-in-Publication Data: A catalogue record for this book is available from the British Library.

Published by IOP Publishing, wholly owned by The Institute of Physics, London

IOP Publishing, No.2 The Distillery, Glassfields, Avon Street, Bristol, BS2 0GR, UK

US Office: IOP Publishing, Inc., 190 North Independence Mall West, Suite 601, Philadelphia, PA 19106, USA

Contents

Preface

We were motivated to write this book by our research findings in the fields of perovskite-based optoelectronic devices and plasmonic devices over the last decade. This book is strongly related to our previous courses, namely 'Light-Emitting Diodes and Solar Cells,' 'Technology for Optical Systems,' and 'Laser Physics and Optical Spectroscopy Techniques.' Ten chapters are used to mathematically and graphically describe light–material interactions, optical spectrometers, and the working mechanisms of optical and optoelectronic devices. This book may play a vital role in multidisciplinary sciences. It is suitable for use as a textbook in graduate courses and as a reference book in undergraduate courses.

The contents of chapters 1–3 are suitable for optics-related courses. The contents of chapter 4 are mainly related to various optical-microscopy-based spectrometers and their related applications in materials characterization, as well as optical analysis for device physics; it is suitable for graduate students in the fields of optics, photonics, materials science, and solid-state physics. In chapter 5, the theory and related applications of optical waveguides are graphically and mathematically described; this chapter is especially suitable for graduate students in the fields of optics and photonics.

The contents of chapters 6–8 are mainly related to the properties and manipulation of excitons and photogenerated carriers in various semiconducting materials and are suitable for graduate students in the fields of photonics, materials science, and solid-state physics. In chapters 9 and 10, the working mechanisms of various solar cells and light-emitting diodes are graphically and conceptually described at a suitable level for graduate students in the fields of photonics, materials science, and solid-state physics.

Overall, this book is suitable for graduate-level courses in colleges of science or electrical engineering. The main purpose of this book is to guide senior undergraduate students and graduate students.

Acknowledgments

Dr Anjali Chandel wishes to express her deep appreciation to her family, friends together with her advisor for their support. She also expresses thanks to the National Science and Technology Council (NSTC) of Taiwan for the financial assistance provided through grant numbers 113-2811-M019-012 and 113-2811-M-033-002. Professor Sheng Hsiung Chang thanks the supports from his familily and the NSTC and Ministry of Science and Technology (grant no. 107-2112-M-033-001-MY3, 110-2112-M-033-008-MY3 and 113-2112-M-019-005). Besides, Professor Chang thanks his previous undergraduate and graduate students for the helpful discussions while preparing this book.

Author biographies

Anjali Chandel

Anjali Chandel was born and raised in Himachal Pradesh, India. She earned a PhD in Physics from Chung Yuan Christian University, Taiwan, based on research focused on perovskite solar cells and their applications. Currently, she is a postdoctoral researcher at National Taiwan Ocean University, Taiwan. Her work aims to advance the efficiency and practical deployment of perovskite-based technologies in renewable energy.

Sheng Hsiung Chang

Sheng Hsiung Chang received his BS degree from the Department of Physics at National Chung Hsing University, Taichung, ROC, in 2002; he received his Master's degree from the Institute of Optical Science at the National Central University (NCU), Taoyuan, ROC, in 2004; and he received his PhD degree from the Department of Electrical Engineering at the NCU, Taoyuan, ROC, in 2008.

From November 2008 to July 2012, he was a postdoctoral research fellow at the Modern Optical Spectroscopy Laboratory of the Institute of Atomic and Molecular Sciences, Academia Sinica, and the Department of Optics and Photonics, NCU. In August 2012, he joined the faculty of the Research Center for New Generation Photovoltaics at the NCU as an assistant research fellow and was promoted to associate research fellow. In March 2018, he joined the faculty of the Department of Physics at Chung Yuan Christian University (CYCU) as an associate professor. He was promoted to professor at CYCU in August 2022. In August 2024, he joined the faculty of the Department of Optoelectronics and Materials Technology at National Taiwan Ocean University as a professor.

Dr Chang is a member of the Phi Tau Phi Scholastic Honor Society (2004), the editorial board of the *Journal of TVS*, and the editorial board of *Nanotechnology* (2019–) and a former member of its advisory panel (2017–19). He was an R&D 100 Awards Winner in 2015. He received an Outstanding Research Award from the NCU in 2016 and 2017. He received the MOST Special Outstanding Talent Award for 2018, 2019, 2020, 2021, and 2022. He received the Publons Peer Review Award 2018 in Nanotechnology and Materials Science. In December 2020, he received the Taiwan Vacuum Society's Young Scientist Award for 2020. He received an Outstanding Research Award from CYCU for 2021. Currently, he is serving on the editorial board of *Nanotechnology*. Dr Chang's research is focused on interfacial contacts and nanophase materials used in the development of highly efficient and stable inverted perovskite solar cells.

List of symbols

α	Polarizability
	Absorption coefficient
ε	Permittivity
η	Wave impedance
λ_0	Wavelength in free space
ρ	Density
χ	Susceptibility
μ	Permeability
	Mobility
σ	Conductivity
τ	Relaxation time
v	Velocity
ω	Angular frequency
θ	Angle
θ_E	Einstein phonon temperature
ϕ	Phase
	Potential
	Wave function
Γ	Collision frequency
Δ	Phase
A	Area
	Amplitude
B	Magnetic field
D	Electric displacement field
	Diffusion coefficient
E	Electric field
	Energy
E_b	Exciton binding energy
E_g	Energy bandgap
H	Magnetic field strength
F	Force
	Focal length
I	Intensity
	Current
J	Current density
K	Spring constant
K_B	Boltzmann constant
N	Carrier density
P	Polarization
R	Reflectance
	Resistance
T	Transmittance
	Period
	Temperature
U	Potential energy
V_d	Abbe number
W	Width

Y	Admittance
a	Acceleration
	Radius
c	Light velocity in free space
e	Elementary charge
h	Planck constant
f	Frequency
k	Propagation constant
m	Mass
n	Refractive index
q	Charge quantity
r	Reflectivity
	Radius
t	Transmittivity
	Thickness
	Time
w	Width

IOP Publishing

Light–Material Interactions and Applications in
Optoelectronic Devices

Anjali Chandel and Sheng Hsiung Chang

Chapter 1

Wave propagation, properties, and manipulation

In this chapter, wave propagation, properties, and manipulation are mathematically described using wave equations to explain light–material interactions. The frequency, wavelength, propagation constant, and wave fronts of an electromagnetic (EM) plane wave are defined by analyzing Maxwell's equations. In regard to wave manipulations, the redirection of propagation, the control of polarization, the formation of interference and diffraction, and the generation of nonlinear optical responses are conceptually illustrated using logical graphic and/or mathematic methods. To help the reader understand EM wave properties, the diffraction pattern of a plane wave through a narrow slit is formulated as a Fourier transform that also describes the relationship between real space and momentum space. On the subject of nonlinear optical responses, second-harmonic generation and the optical Kerr effect are described to help the reader fully understand light–material interactions.

1.1 Maxwell's equations

To understand the propagation, properties, and manipulation of electromagnetic (EM) waves in materials, it is beneficial to study Maxwell's equations, which describe the interactions between EM waves and charged particles in materials. From a mathematic point of view, Maxwell's equations describe the divergence and curl of electric fields (E) and magnetic fields (B) in materials. In a uniform space, a radial electric field can be generated by an electrically charged particle; such an electric field can be described using the electric Gauss's law shown in equation (1.1a). If a magnetic monopole exists in a uniform space, a radial magnetic field can be generated, which is described by the magnetic Gauss's law shown in equation (1.1b). Time-varying magnetic fields and magnetic currents can generate a curl in electric fields, which is described using Faraday's law of induction shown in equation (1.1c). Time-varying electric fields and electric currents can also generate a curl in magnetic

doi:10.1088/978-0-7503-6099-9ch1

fields, which is described by Ampère's circuital law shown in equation (1.1d). It should be noted that equations (1.1a), (1.1b), (1.1c), and (1.1d) are the derived form in the time domain.

$$\nabla \cdot (\varepsilon \vec{E}) = \rho_e \tag{1.1a}$$

$$\nabla \cdot \vec{B} = \rho_m \tag{1.1b}$$

$$\nabla \times \vec{E} = -\frac{\partial \vec{B}}{\partial t} - \vec{J_m} \tag{1.1c}$$

$$\nabla \times \left(\frac{1}{\mu} \vec{B} \right) = \frac{\partial (\varepsilon \vec{E})}{\partial t} + \vec{J_e} \tag{1.1d}$$

where ε and μ are the permittivity and permeability of the uniform environment, respectively; ρ_e and ρ_m are the densities of the electrically charged particles and magnetic monopoles, respectively; and J_e and J_m are the densities of the electric current and magnetic current, respectively. Up until now, magnetic monopoles have not been found because magnetic field lines naturally form closed loops. When two particles of opposite electric charge are placed in a uniform and isotropic space, the spatial distribution of the resulting electric field can be described using Coulomb's law, which can be used to calculate the coulomb energy between an electron and a hole. Conceptually, coulomb energy is similar to exciton (electron–hole pair) binding energy, which is inversely proportional to the permittivity of the space and the distance between the paired electron and hole [1]. In other words, Gauss's law can be used to understand the physical meaning of the binding energy between electron and hole clouds in materials. On the other hand, equations (1.1c) and (1.1d) show that alternating current (AC) sources generate time-varying electric and magnetic fields, thereby forming EM waves. When J_e equals $J_{e0} \times (\sin\omega t)$ in equation (1.1d), the induced magnetic field is a time-varying vector field that generates the time-varying electric field in equation (1.1c). When J_m equals $J_{m0} \times (\sin\omega t)$ in equation (1.1c), the induced electric field is a time-varying vector field that generates the time-varying magnetic field in equation (1.1d). In other words, Faraday's law of induction and Ampère's circuital law can be used to explain the EM waves radiated by AC sources. Figure 1.1 displays the instantaneous E_z distributions of a cylindrical EM wave radiated from an infinitely long and harmonic current source in the z-direction. Over time, the EM wave propagates from the center to the four boundaries. At $t = $ T6, the cylindrical EM wave is not reflected from the boundaries owing to the use of perfectly matched layers (PMLs). The radially outgoing EM wave is effectively absorbed by the PMLs, thereby avoiding the formation of interference patterns in the simulation region, which can be computed using the finite-difference time-domain (FDTD) method [2]. The area of the simulation space is about $4\lambda \times 4\lambda$ when described in terms of the wavelength λ.

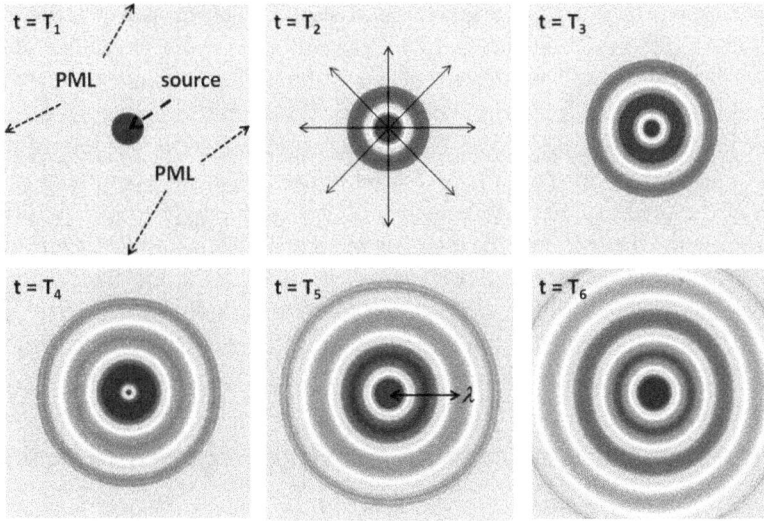

Figure 1.1. E_z distribution of a cylindrical EM wave at the different moments: T_1, T_2, T_3, T_4, T_5, and T_6.

1.2 Wave propagation and properties

A plane wave equation can be deduced by taking the curl of equation (1.1c) or (1.1d) when the AC sources are removed. Equation (1.2a) is the curl of equation (1.1c) when the magnetic current density is zero. In free space, the density of electrically charged particles is zero in equation (1.1a), which results in the zero divergence of electric fields in space. After inserting equation (1.1d) into equation (1.2a), the space-related and time-varying electric fields can be written as equation (1.2b). When the environment is uniform and isotropic, the equation for electric fields can be written as equation (1.2c), which is the plane wave equation in the form of an electric field. To solve for the propagation velocity of EM waves, the vector form of the electric field is given in equation (1.2d), which is the plane wave function in the form of an electric field. The vectors k and ω are the propagation constant and the angular frequency of the plane wave, respectively. The vector form of the magnetic field is given in equation (1.2e). After inserting the vector form of the electric field into equation (1.2c), the equation for electric fields can be written as equation (1.2f). Because the electric fields are nonzero, the relation between the wave properties (k and ω) and the material properties (μ and ε) can be determined using equation (1.2f). In other words, the dispersion relation ($\omega^2/k^2 = 1/(\mu\varepsilon)$) is obtained. In a harmonic EM wave, k and ω equal $2\pi/\lambda$ and $2\pi f$, respectively, where λ is the wavelength of the harmonic EM wave. Therefore, the dispersion relation can be rewritten as $f\lambda = 1/(\mu\varepsilon)^{1/2} = v$, where v is defined as the propagation velocity of the harmonic EM wave in a uniform and isotropic space. In a vacuum environment, the dispersion relation can be written as $f\lambda_0 = 1/(\mu_0\varepsilon_0)^{1/2} = c$, where c is the propagation velocity of the harmonic EM wave. Here, μ_0 and ε_0 are the absolute permittivity (8.854×10^{-12} F m^{-1}) and the absolute permeability ($4\pi \times 10^{-7}$ H m^{-1}), respectively. F and H denote units of faradays and henries, respectively. Therefore, the calculated c value equals 2.9979×10^8 m s^{-1}, which shows that vacuum is a dispersionless space. The ratio of λ to λ_0 is defined as the refractive index (n) of a uniform and isotropic environment. In a source-free, uniform, and isotropic environment, the directions of the electric field, magnetic field, and EM wave propagation can be determined using equations (1.1c), (1.2d), and (1.2e). After taking the curl of the vector E and the time derivative of the vector B, the new vector function can be written as equation (1.2g). In other words, the direction of the cross product of vectors k and E is parallel to the vector B. When the directions of the vectors E and B lie along the x- and y-axes, respectively, the direction of the vector k lies along the z-axis, as shown in figure 1.2. The distance between two adjacent wave fronts is defined as one wavelength of the harmonic EM wave in the medium.

$$\nabla \times \nabla \times \overrightarrow{E} = \nabla (\nabla \cdot \overrightarrow{E}) - \nabla^2 \overrightarrow{E} = \nabla \times \left(-\frac{\partial \overrightarrow{B}}{\partial t} \right) = -\frac{\partial}{\partial t}(\nabla \times \overrightarrow{B}) \quad (1.2a)$$

$$-\nabla^2 \overrightarrow{E} = -\frac{\partial}{\partial t}(\nabla \times \overrightarrow{B}) = -\frac{\partial}{\partial t}\left[\mu \frac{\partial}{\partial t}(\varepsilon \overrightarrow{E}) \right] \quad (1.2b)$$

$$\nabla^2 \overrightarrow{E} - \mu\varepsilon \frac{\partial^2}{\partial t^2}(\overrightarrow{E}) = 0 \quad (1.2c)$$

$$\overrightarrow{E} = \overrightarrow{E_0} \exp[i(\overrightarrow{k} \cdot \overrightarrow{r} - \omega t)] \quad (1.2d)$$

$$\overrightarrow{B} = \overrightarrow{B_0} \exp[i(\overrightarrow{k} \cdot \overrightarrow{r} - \omega t)] \quad (1.2e)$$

$$(-k^2 + \mu\varepsilon\omega^2)\overrightarrow{E} = 0 \quad (1.2f)$$

$$\overrightarrow{k} \times \overrightarrow{E} = \omega\overrightarrow{B} \quad (1.2g)$$

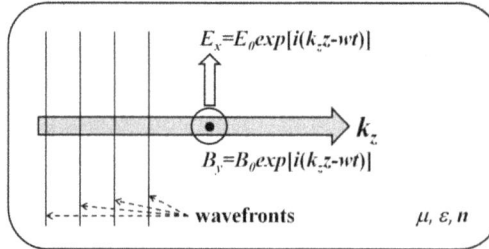

Figure 1.2. The wave fronts, electric field (E_x), magnetic field (B_y), and propagation constant (k_z) of an EM plane wave in a uniform and isotropic environment.

1.3 Wave manipulation

In linear optics, the propagation direction and properties of a plane wave are changed at an interface, as shown in figure 1.3, which can be described using Snell's law. At the n_H/n_L interface (see figure 1.3(a)), the propagation constants k_i, k_r, and k_t equal $n_H k_0$, $n_H k_0$, and $n_L k_0$, respectively, where n_H and n_L are the refractive index values in the top and bottom regions, respectively. k_0 is the wave propagation constant of free space. The subscripts i, r, and t denote incidence, reflection, and transmission, respectively. Snell's law can be used to calculate the reflection angle θ_r and transmission angle θ_t when the incident angle θ_i is known. The formulas of Snell's law for refraction and reflection are written in equations (1.3a) and (1.3b), respectively. Here, n_i and θ_i are equal to n_{inc} and θ_{inc}, respectively. θ_r always equals θ_i in a uniform medium. θ_t is proportional to θ_i and n_i/n_t. When θ_t equals $\pi/2$, θ_i is defined as the critical angle θ_c ($= \sin^{-1}(n_t/n_i)$). Total internal reflection (TIR) occurs at the n_H/n_L interface when θ_i is larger than θ_c. In a GaN-based light-emitting diode, the refractive index values of the GaN substrate ($n_i = n_H$) and air ($n_t = n_L$) are about 3.4 and 1.0 in the visible wavelength range, respectively. This results in a small critical angle of 17.1° and thereby naturally limits the light extraction efficiency. At the n_L/n_H interface (see figure 1.3(b)), θ_t is always smaller than θ_i. Therefore, the TIR phenomenon cannot occur at the n_L/n_H interface when EM waves pass from a low-index medium to a high-index medium. When the incident angle θ_i equals zero, the reflectance R and transmittance T can be calculated using equations (1.3c) and (1.4c). In highly efficient perovskite solar cells, glass plates are widely used as the substrate; these have a low refractive index, thereby resulting in low reflection at normal incidence. The calculated reflectance at an air/glass interface is about 3.37% when the refractive index of glass is about 1.45 in the visible wavelength range. When the refractive index of the substrate is increased to 4, the calculated reflectance value is 36%. Equation (1.3c) shows that the reflectance value increases as the refractive index increases. Equation (1.3c) is still valid when the refractive index of the substrate is a complex number. For example, the refractive index of Ge is 4.3935 + i2.3981 at a wavelength of 500 nm, which results in a reflectance value of 49.56% at an air/Ge interface. The high reflectance value at the air/Ge interface explains why Ge wafers look like mirrors.

$$n_{inc} \sin \theta_{inc} = n_t \sin \theta_t \qquad (1.3a)$$

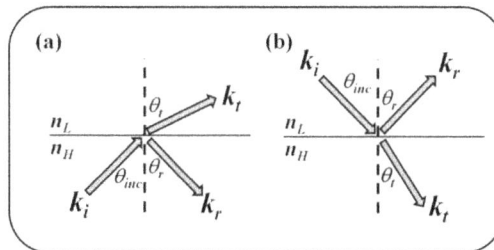

Figure 1.3. Reflection and transmission of a plane wave at an interface. (a) n_H/n_L interface; (b) n_L/n_H interface.

$$n_{inc} \sin \theta_{inc} = n_{inc} \sin \theta_r \qquad (1.3b)$$

$$R = |(n_L - n_H)/(n_L + n_H)|^2 \qquad (1.3c)$$

$$T = 1 - R \qquad (1.3d)$$

The properties of EM waves can also be manipulated by varying polarization states, interference effects, diffraction effects, and nonlinear optical responses, which are described in the following subsections.

1.3.1 Polarization

A polarized EM wave can be described by the vectors E and H, whose directions may be given at any moment in time. The polarizations of light waves are classified into linear, circular, and elliptical polarizations, which can be described using Jones matrices [3]. Their Jones matrices are listed in figures 1.4(a), (c), and (d). The electric field of a plane wave propagating in the z-direction can be decomposed into E_x and E_y in cartesian coordinates, as shown in figure 1.4(b). The angle θ determines the magnitudes of E_x and E_y. The Jones matrix for linear polarization is presented in figure 1.4(a). When linear polarization is present, E_x and E_y are in phase. When the θ value equals zero, the E_x and E_y values associated with linear polarization are zero and one, respectively. The Jones matrices of right-handed and left-handed elliptical polarizations are presented in figures 1.4(c) and (d), respectively. When the phase difference between E_x and E_y equals $\pi/2$ ($-\pi/2$), the polarization is defined as right-handed (left-handed) elliptical polarization, where the sense of rotation of the polarization plane is clockwise (anticlockwise). When the angle θ equals $\pi/4$, right-handed (left-handed) elliptical polarization reduces to right-handed (left-handed) circular polarization. Linear polarizers, polarization beam splitters, half-wave plates, and quarter-wave plates can be used to vary the polarization state of plane waves. Their Jones matrices are listed in figures 1.5(a)–(d). Besides, the rotaion matrix and the inverse of the ratation matrix are listed in figures 1.5(e) and 1.5(f), resepctively. The relationship between the polarization states of an incident plane

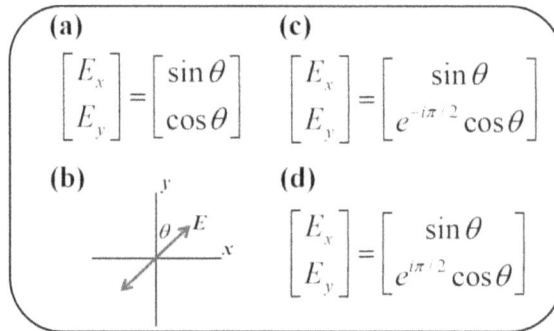

Figure 1.4. Jones matrices and an electric field in cartesian coordinates. (a) Linear polarization; (b) linearly polarized electric field; (c) right-handed elliptical polarization; and (d) left-handed elliptical polarization.

$$
\begin{array}{lll}
\textbf{(a)} & \textbf{(c)} & \textbf{(e)} \\[6pt]
LP_x = \begin{bmatrix} 1 & 0 \\ 0 & 0 \end{bmatrix} & WP_{1,2} = \begin{bmatrix} 1 & 0 \\ 0 & -1 \end{bmatrix} & R = \begin{bmatrix} \cos\theta & -\sin\theta \\ \sin\theta & \cos\theta \end{bmatrix} \\[12pt]
\textbf{(b)} & \textbf{(d)} & \textbf{(f)} \\[6pt]
LP_y = \begin{bmatrix} 0 & 0 \\ 0 & 1 \end{bmatrix} & WP_{1,4} = \begin{bmatrix} 1 & 0 \\ 0 & i \end{bmatrix} & R^{-1} = \begin{bmatrix} \cos\theta & \sin\theta \\ -\sin\theta & \cos\theta \end{bmatrix}
\end{array}
$$

Figure 1.5. Jones matrices: (a) x-polarizer; (b) y-polarizer; (c) half-wave plate; (d) quarter-wave plate; (e) rotation matrix; and (f) inverse of rotation matrix.

wave and a transmitted plane wave can be obtained using equations (1.4a), (1.4b), and (1.4c), where R is a rotation matrix and R^{-1} is the inverse of the rotation matrix. In equation 1.4(a), the θ value determines the magnitude of the transmitted plane wave. The transmittance values are zero and one when the θ values of the linear polarizer are zero and $\pi/2$, respectively. This shows that a linear polarizer can be used as an on–off switch by rotating it through an angle of θ. In equation 1.4(b), E_x and E_y can be expressed as $\cos\theta\sin\theta$ and $\sin^2\theta - \cos^2\theta$, respectively. The E_x and E_y values range from -1 to 1, which shows that a half-wave plate can be used to arbitrarily rotate the direction of an electric field in the x–y plane when a plane wave propagates in the z-direction. In equation 1.4(c), E_x and E_y can be expressed as $(1 - i)\sin\theta\cos\theta$ and $\sin^2\theta + i\cos^2\theta$, respectively. When the θ value of a quarter-wave plate equals $\pi/4$, E_x and E_y equal $(1-i)/2$ and $(1 + i)/2$, respectively, which results in a 1×2 matrix as written in equation 1.4(d). This shows that a linearly polarized plane wave can be converted into a circularly polarized plane wave by passing through a $\pi/4$-oriented quarter-wave plate.

It should be noted that the half-wave and quarter-wave plates are lossless biaxial crystals. In other words, the transmittance of these wave plates equals unity when the reflections at the two interfaces are ignored. Therefore, a normalization process must be applied in order to satisfy energy conservation, as shown in figure 1.5(e). For example, the angle-dependent output 1×2 matrix of a half-wave plate can be normalized using equation (1.4f).

$$
\begin{bmatrix} E_x \\ E_y \end{bmatrix} = R\begin{bmatrix} 1 & 0 \\ 0 & 0 \end{bmatrix}R^{-1}\begin{bmatrix} 0 \\ 1 \end{bmatrix} = \begin{bmatrix} \cos^2\theta & \cos\theta\sin\theta \\ \cos\theta\sin\theta & \sin^2\theta \end{bmatrix}\begin{bmatrix} 0 \\ 1 \end{bmatrix} \tag{1.4a}
$$

$$
\begin{bmatrix} E_x \\ E_y \end{bmatrix} = R\begin{bmatrix} 1 & 0 \\ 0 & -1 \end{bmatrix}R^{-1}\begin{bmatrix} 0 \\ 1 \end{bmatrix} = \begin{bmatrix} \cos^2\theta - \sin^2\theta & \cos\theta\sin\theta \\ \cos\theta\sin\theta & \sin^2\theta - \cos^2\theta \end{bmatrix}\begin{bmatrix} 0 \\ 1 \end{bmatrix} \tag{1.4b}
$$

$$
\begin{bmatrix} E_x \\ E_y \end{bmatrix} = R\begin{bmatrix} 1 & 0 \\ 0 & i \end{bmatrix}R^{-1}\begin{bmatrix} 0 \\ 1 \end{bmatrix} = \begin{bmatrix} \cos^2\theta + i\sin^2\theta & (1 - i)\sin\theta\cos\theta \\ (1 - i)\sin\theta\cos\theta & \sin^2\theta + i\cos^2\theta \end{bmatrix}\begin{bmatrix} 0 \\ 1 \end{bmatrix} \tag{1.4c}
$$

$$
\begin{bmatrix} E_x \\ E_y \end{bmatrix} = \frac{1}{2}\begin{bmatrix} 1 - i \\ 1 + i \end{bmatrix} = \frac{\sqrt{2}}{2}\begin{bmatrix} e^{-i\pi/4} \\ e^{i\pi/4} \end{bmatrix} = \frac{\sqrt{2}}{2}e^{-i\pi/4}\begin{bmatrix} 1 \\ e^{i\pi/2} \end{bmatrix} = \frac{\sqrt{2}}{2}e^{-i\pi/4}\begin{bmatrix} 1 \\ i \end{bmatrix} \tag{1.4d}
$$

$$\begin{bmatrix} E_x \\ E_y \end{bmatrix} \xrightarrow{\text{Nor.}} \frac{1}{\sqrt{E_x^2 + E_y^2}} \begin{bmatrix} E_x \\ E_y \end{bmatrix} \tag{1.4e}$$

$$\begin{bmatrix} \cos\theta \sin\theta \\ \sin^2\theta - \cos^2\theta \end{bmatrix} \xrightarrow{\text{Nor.}} \frac{1}{\sqrt{(\cos\theta \sin\theta)^2 + (\sin^2\theta - \cos^2\theta)^2}} \begin{bmatrix} \cos\theta \sin\theta \\ \sin^2\theta - \cos^2\theta \end{bmatrix} \tag{1.4f}$$

1.3.2 Interference

The electric fields of plane waves can be expressed as sinusoidal functions. When two E_y-directed plane waves propagate in the z-direction, the total electric field can be written using equations (1.5a) and (1.5b), where k, ω, and ϕ are the propagation constant, angular frequency, and initial phase of the plane waves. The subscripts A and B denote the two plane waves. When the plane waves A and B have the same angular frequency, i.e. $\omega_A = \omega_B$, the propagation constants k_A and k_B are also the same. Then, the total electric field reduces to equation (1.5c). The $E_{A,y}e^{i\phi_A} + E_{B,y}e^{i\phi_B}$ term dominates in determining the interference result. When the plane waves A and B have the same initial phase, i.e. $\phi_A = \phi_B$, the total electric field equals the summation of $E_{A,y}$ and $E_{B,y}$, which is defined as constructive interference. When plane waves A and B are out of phase, i.e. $\phi_A - \phi_B = \pi(2n + 1)/2$, where n is an integer, the total electric field equals the difference between $E_{A,y}$ and $E_{B,y}$, which is defined as destructive interference. The working mechanism of a Mach–Zehnder interferometer [4, 5] can be understood using equation (1.5c). Figure 1.6 presents a waveguide-based Mach–Zehnder interferometer. The propagation properties of a waveguide mode and a plane wave are similar. Therefore, the interference at the Y-combiner of the Mach–Zehnder interferometer can be expressed using equation (1.5c) when the plane waves A and B are assigned to the waveguide modes of the top arm and bottom arm, respectively. When the Y-splitter and Y-combiner are symmetric waveguide structures, the magnitudes of $E_{A,y}$ and $E_{B,y}$ are the same. When the refractive index of the top arm is increased through the use of electro-optic (EO) effects, the phase difference between the two modes corresponding to the top and bottom arms is also increased. In other words, the interference at the

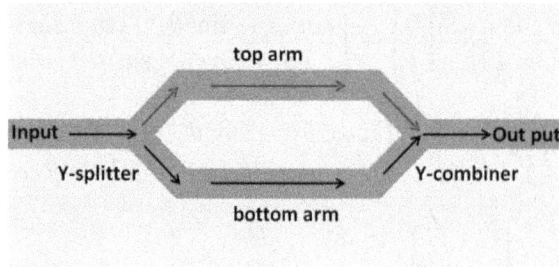

Figure 1.6. The waveguide configuration of a Mach–Zehnder interferometer.

Y-combiner of a Mach–Zehnder interferometer can be manipulated by varying the refractive index of the top arm and/or bottom arm. Therefore, the Mach–Zehnder interferometer can be used as a fast on–off switch due to the fast EO effects.

$$E_y(t) = E_{A,y}\cos(k_A z - \omega_A t + \phi_A) + E_{B,y}\cos(k_B z - \omega_B t + \phi_B) \qquad (1.5a)$$

$$\text{Re}[E_y(t)] = \text{Re}[E_{A,y}e^{i(k_A z - \omega_A t + \phi_A)} + E_{B,y}e^{i(k_B z - \omega_B t + \phi_B)}] \qquad (1.5b)$$

$$\text{Re}[E_y(t)] = \text{Re}[e^{i(kz - \omega t)}(E_{A,y}e^{i\phi_A} + E_{B,y}e^{i\phi_B})] \qquad (1.5c)$$

1.3.3 Diffraction

The geometric shape of the wave front of an EM wave stays the same as it propagates through a homogeneous and uniform medium. The wave front and intensity distribution change when an EM wave travels through an inhomogeneous medium, which may contain opaque diaphragms or regions with considerable fluctuations in refractive index. The phenomenon that takes place under the latter condition is called diffraction; it originates from the structure-induced change in the wave front of the EM wave.

In the linear propagation of a plane wave, the infinite wave front explains the diffractionless propagation characteristics in a uniform medium. However, wave diffraction can be demonstrated by observing the transmitted pattern of a plane wave through a narrow slit, as shown in figure 1.7. When a wave passes through a slit, its spatial distribution becomes rectangular. As a result of wave diffraction, the intensity distribution on the screen takes the form of a squared sinc function. Theoretical predictions show that the distribution of the diffraction pattern is wider when the width of the slit is narrower. In other words, a narrower slit results in a

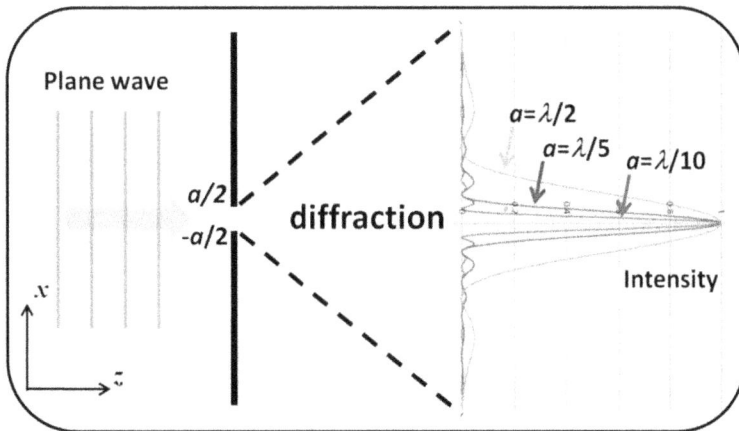

Figure 1.7. Diffraction of an EM wave though a single slit.

broader distribution of the wave vector. The diffraction pattern of a single slit can be obtained by solving the Fourier transform of the rectangular function given in equation (1.6a), where a is the width of the slit. After using the relation ($k = 2\pi(\sin\theta)/\lambda$), the function of k can be changed to a function of θ, as written in equation (1.6b), where α is defined as $a\pi(\sin\theta)/\lambda$ and λ is the wavelength of the plane wave. Equation (1.6c) shows that the diffraction pattern is the square of a sinc function.

$$\tilde{E}(k) = \frac{1}{a}\int_{-a/2}^{a/2} 1 \times e^{-ikx}dx = \frac{1}{a}\left(\frac{-1}{ik}e^{-ikx}\Big|_{-a/2}^{a/2}\right) = [\sin(ak/2)]/(ak/2) \quad (1.6a)$$

$$\tilde{E}(\theta) = \left[\sin\left(\frac{a}{2}\frac{2\pi}{\lambda}\sin\theta\right)\right]\Big/\left[\frac{a}{2}\frac{2\pi}{\lambda}\sin\theta\right] = \frac{\sin\alpha}{\alpha} \quad (1.6b)$$

$$I = |\tilde{E}(\theta)|^2 = \frac{\sin^2\alpha}{\alpha^2} \quad (1.6c)$$

1.3.4 Nonlinear optical responses

Nonlinear optical phenomena depend on the light intensity in materials. The dielectric responses of an isotropic and uniform material can be described by using the dielectric constitutive relation which is written in equation (1.7a), where vector D is electric displacement and vector P is electric dipole moment. The interaction between the EM wave and material induces the electric dipole moment, which can be expressed by using equation (1.7b). The first term $\varepsilon_0\chi^{(1)}\vec{E}(t)$ is linearly proportional to the electric field, which is responsible for the linear optical responses. The second term $\varepsilon_0\chi^{(2)}\vec{E}^2(t)$ and third term $\varepsilon_0\chi^{(3)}\vec{E}^3(t)$ are related to the second harmonic generation and optical Kerr effect (OKE), respectively. When the electric field of the incident EM wave is along one of the crystal axes, the ith component of electric dipole moment (P_i) can be written as equation (1.7c).

$$\vec{D} = \varepsilon\vec{E} = \varepsilon_0\vec{E} + \vec{P} \quad (1.7a)$$

$$\vec{P}(t) = \varepsilon_0\left(\chi^{(1)}\vec{E}(t) + \chi^{(2)}\vec{E}^2(t) + \chi^{(3)}\vec{E}^3(t) + \cdots\right) \quad (1.7b)$$

$$P_i = \varepsilon_0\left(\sum_{j=1}^{3}\chi^{(1)}_{ij}E_j + \sum_{j=1}^{3}\sum_{k=1}^{3}\chi^{(2)}_{ijk}E_jE_k + \sum_{j=1}^{3}\sum_{k=1}^{3}\sum_{l=1}^{3}\chi^{(3)}_{ijkl}E_jE_kE_l + \cdots\right) \quad (1.7c)$$

In the second-order nonlinear optical response, the relation between the $P^{2\omega}$ and E can be written in equation (1.7d), which shows that the generation of a second harmonic lightwave. The generation efficiency of the second harmonic lightwave is proportional to the second-order nonlinear-optical coefficient ($\chi^{(2)}$) and the light intensity (E^2). It is noted that the $\chi^{(2)}$ value in a symmetric crystal structure equals to zero because the induced $P^{2\omega}$ values are positive and negative in the positive and negative half cycles of the incident EM wave, respectively. In other words, the crystal with a polar structure can be used as the second harmonic generator, such as beta barium borate (β-BBO) and lithium triborate (LBO) and potassium titanyl phosphate (KTP). β-BBO crystals are widely used to efficiently convert the near-infrared femtosecond pulsed lasers to the visible femteosecond pulsed lasers. When the phase mismatch between the fundamental wave and the second harmonic wave is considered, the second harmonic generation (SHG) efficiency is dependent on the interaction length, which is expressed in equation (1.7e). $k_{2\omega}$ and k_ω are propagation constants of the second harmonic wave and fundamental wave, respectively. To simplify the expression, $\Delta k = k_{2\omega} - k_\omega$ is used. In general, $k_{2\omega}$ is larger than k_ω because the operating frequencices are in the positive dispersion range of the nonlinear crystals, i.e. $dn/d\omega > 1$. With the initial condition ($E(2\omega, z = 0) = 0$), the electric field of the second harmonic wave ($E(2\omega)$) is a function of l, which can be expressed in equation (1.7f), where l is the thickness of the SHG crystal plate. When the relation between intensity and electric field ($I = (n/2)\sqrt{\varepsilon_0/\mu_0}\,|E|^2$) is used, the SHG intensity can be expressed in equation (1.7g). When the phase mismatch can be effectively reduced ($\Delta k \sim 0$), the SHG intensity is linearly proportional to the square of excitation intensity ($I^2(\omega)$) and the square of crystal thickness (l^2).

$$P^{2\omega} = \frac{\varepsilon_0}{2}\chi^{(2)}E^2(e^{j2\omega t} + 2e^0 + e^{-j2\omega t}) = \varepsilon_0\chi^{(2)}E^2[1 + 1\cos(2\omega t)] \quad (1.7\text{d})$$

$$\frac{\partial E(2\omega)}{\partial z} = -\frac{i\omega}{2n_{2\omega}c}\chi^{(2)}E^2(\omega)e^{i(k_{2\omega}-k_\omega)z} \quad (1.7\text{e})$$

$$\int_0^l dE(2\omega) = E(2\omega, l) = -\frac{i\omega\chi^{(2)}E^2(\omega)}{2n_{2\omega}c}\int_0^l e^{i\Delta kz}dz = \frac{\omega\chi^{(2)}E^2(\omega)}{2n_{2\omega}c(\Delta k)}(1 - e^{i\Delta kl}) \quad (1.7\text{f})$$

$$I_{SHG} = I(2\omega, z = l) = \frac{1}{2n_{2\omega}c^3\varepsilon_0}\left(\frac{\omega\chi^{(2)}I(\omega)l}{n_\omega}\right)^2 \frac{\sin^2(\Delta kl/2)}{(\Delta kl/2)^2} \quad (1.7\text{g})$$

In the third-order nonlinear optical response, the OKE induced electrical dipole moment can be written as equation (1.7h). The electric field and propagation of the incident EM wave are along the x-axis and z-axis, respectively. The average refractive index of the OKE materials can be written as equation (1.7i). In general, the third-order nonlinear optical response is weak. Therefore, the OKE induced change in the refractive index can be expressed in equation (1.7j), which shows that the change in the refractive index is proportional to the intensity of the incident EM wave. When a femtosecond pulsed laser beam propagates in an OKE material, the refractive index in the central region is larger than that in the outer region and thereby forming the self-focusing of a Gaussian lightwave which increases the light-material interaction strength.

$$P^{OKE}(t) = \varepsilon_0 E_x(\chi^{(1)} + 3\chi^{(3)}|E_x|^2\cos^2(t)) = (\varepsilon_0 E_x)(\varepsilon_r) \quad (1.7\text{h})$$

$$n = (n_0^2 + 3\chi^{(3)} |E_x|^2 \int_0^T \cos^2(\omega t) dt)^{1/2} = (n_0^2 + \frac{3}{2}\chi^{(3)} |E_x|^2 t)^{1/2} \qquad (1.7i)$$

$$n \sim n_0 + \frac{3\chi^{(3)}}{4} |E_x|^2 = n_0 + n_2 I \qquad (1.7j)$$

Bibliography

[1] Nayak P K 2013 Exciton binding energy in small organic conjugated molecule *Synth. Met.* **174** 42–5

[2] Taflove A and Hagness S C 2000 *Computational Electrodynamics: The Finite-Difference Time-Domain Method* 2nd edn (Norwood, MA: Artech House)

[3] Jones R C 1941 New calculus for the treatment of optical systems *J. Opt. Soc. Am* **31** 488–93

[4] Zehnder L 1891 Ein neuer Interferenzrefraktor *Z. Instrumk.* **11** 275–85

[5] Mach L 1892 Ueber einen Interferenzrefraktor *Z. Instrumk.* **12** 89–93

IOP Publishing

Light–Material Interactions and Applications in Optoelectronic Devices

Anjali Chandel and Sheng Hsiung Chang

Chapter 2

Dielectric responses

In this chapter, anisotropic and dispersive characteristics are mathematically described in order to explain complicated light–material interactions. The constitutive relation between the electric displacement field (D) and the electric field (E) is presented in matrix form. In other words, a dielectric permittivity matrix can be used to relate the D and E vectors, thereby representing the anisotropic characteristics of the material. In the time domain, correlation between a time-varying electric field ($E(t)$) and a time-dependent dielectric permittivity ($\varepsilon(t)$) results in a frequency-dependent dielectric response. From radio frequencies to optical frequencies, dispersive dielectric responses are illustrated using the Debye model, the Lorentz model, and the Drude model. The dispersive dielectric response can be used to predict the phase delay imposed on transmitted EM waves and the absorption of the material, thereby resulting in unique features in the transmittance spectrum which can be analyzed using a transfer matrix method.

2.1 Anisotropic and dispersive characteristics

The electric constitutive relation ($\overrightarrow{D} = \varepsilon\overrightarrow{E}$) is used to describe the relation between an electric field and charged particles. When the material is an anisotropic crystal and an EM wave propagates in the z-direction, the electric constitutive relation can be written in matrix form as equation (2.1a). In general, the diagonal elements ε_{xx} and ε_{yy} correspond to the slow and fast axes, respectively. In addition, the non-diagonal elements are zero in biaxial crystals. When E_y is not aligned with the fast axis of a biaxial crystal, the electric constitutive relation can be written as equation (2.1b), where θ is the angle between E_y of the incident EM wave and the fast axis of the biaxial crystal.

$$\begin{bmatrix} D_x \\ D_y \end{bmatrix} = \begin{bmatrix} \varepsilon_{xx} & \varepsilon_{xy} \\ \varepsilon_{yx} & \varepsilon_{yy} \end{bmatrix} \begin{bmatrix} E_x \\ E_y \end{bmatrix} \quad\quad (2.1a)$$

doi:10.1088/978-0-7503-6099-9ch2

$$\begin{bmatrix} D_x \\ D_y \end{bmatrix} = \begin{bmatrix} \varepsilon_{xx}\cos^2\theta + \varepsilon_{yy}\sin^2\theta & (\varepsilon_{xx} - \varepsilon_{yy})\cos\theta\sin\theta \\ (\varepsilon_{xx} - \varepsilon_{yy})\cos\theta\sin\theta & \varepsilon_{xx}\sin^2\theta + \varepsilon_{yy}\cos^2\theta \end{bmatrix} \begin{bmatrix} E_x \\ E_y \end{bmatrix} \tag{2.1b}$$

In equation (2.1b), the anisotropic and time-averaged dielectric response is illustrated using an angle-dependent 2 × 2 matrix, which does not describe the dispersive properties of materials. The light–material interaction is naturally dispersive owing to the incoherent relation between the driving frequency of the incident EM wave and the oscillation characteristics of the charged particles, as shown in figure 2.1. In the time domain, the dielectric constitutive relation can be written as equation (2.2a), where ε_∞ is the relative permittivity at extremely high frequencies, i.e. $\omega \to \infty$. The dispersion is due to the correlation between the time-varying electric field ($E(t)$) and the susceptibility ($\chi(t)$) of the material. To solve the dielectric constitutive relation in the time domain, the time-varying susceptibility function can be obtained by computing the Fourier transform of the frequency-dependent susceptibility function ($\tilde{\chi}(\omega)$) written in equation (2.2b).

$$D(t) = \varepsilon_0\varepsilon_\infty E(t) + \varepsilon_0 \int_{-\infty}^{t} E(\tau)\chi(t-\tau)\mathrm{d}\tau \tag{2.2a}$$

$$\chi(t) = \frac{1}{2\pi} \int_{-\infty}^{\infty} \tilde{\chi}(\omega)e^{i\omega t}\mathrm{d}\omega \tag{2.2b}$$

The induced electric dipole moment indicates that the propagation of EM waves in the material is retarded; at the macroscopic scale, this behavior can be described using the refractive index. In a non-magnetic material ($\mu_r = 1$), the relation between the relative permittivity and the refractive index can be written as equations (2.3a), (2.3b), and (2.3c). When ε_I is zero, n_R^2 and n_I^2 equal ε_R and zero, respectively. When ε_R is zero, n_R^2 and n_I^2 both equal $\varepsilon_I/2$. The electric field of a plane wave is a function of the refractive index, which can be expressed as equation (2.3d). The intensity decays in the propagation direction (z-direction), which can be expressed by

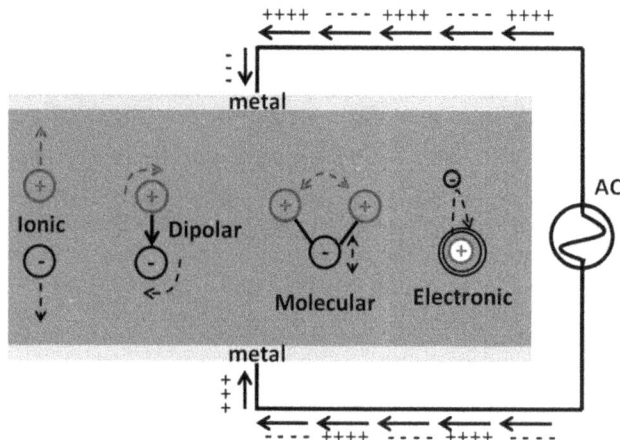

Figure 2.1. Dielectric responses of a multi-property material with an alternating current (AC) source.

equation (2.3e), where the absorption coefficient (α) equals $-2k_0n_I$, and I_0 is the intensity at $z = 0$. Equations (2.3c), (2.3d), and (2.3e) show that the absorption coefficient is related to the frequency of the EM wave and the relative permittivity (ε_R and ε_I). Higher frequencies result in larger absorption coefficients, thereby reducing the propagation lengths of EM waves.

$$\varepsilon_R + i\varepsilon_I = (n_R + in_I)^2 \tag{2.3a}$$

$$n_R^2 = (\varepsilon_R + \sqrt{\varepsilon_R^2 + \varepsilon_I^2})/2 \tag{2.3b}$$

$$n_I^2 = (\sqrt{\varepsilon_R^2 + \varepsilon_I^2} - \varepsilon_R)/2 \tag{2.3c}$$

$$E(t) = E_0 e^{i[k_0(n_R + in_I)z - \omega t]} = E_0 e^{-k_0 n_I z} e^{i(k_0 n_R z - \omega t)} \tag{2.3d}$$

$$I = I_0 e^{-\alpha z} \tag{2.3e}$$

The dielectric responses of materials to an alternating current (AC) source can be classified into ionic diffusion, dipolar reorientation, molecular vibration, and electronic transitions, which are illustrated in figure 2.1. To characterize the dielectric responses of materials from low frequencies to optical frequencies, the time-varying and/or frequency-dependent dielectric functions will be derived in the following sections.

2.2 Dielectric Debye model

When a low-frequency AC source is applied to an insulator with two parallel metallic electrodes (see figure 2.1), the main dielectric responses are related to ionic diffusion and dipolar reorientation. The relaxation time of ionic diffusion is longer than that of dipolar reorientation because of the slow speed of the diffusion process. When the AC source is off, the insulator does not contain a net dipole, because its dipoles are randomly distributed. When the AC source is on, the dipoles in the insulator are aligned with the time-varying electric field between the two parallel metallic electrodes. Conceptually, there is a relaxation time between the electrical excitation and the appearance of an induced electrical dipole moment in the insulator, which results in a dispersive dielectric response. In other words, the electric dipole moment decreases to zero in the form of an exponential decay function, which is written in equation (2.4a), where P_0 is the initially induced electric dipole moment and τ is the relaxation time. The frequency-dependent electric dipole moment can be obtained by computing the Fourier transform of $P(t)$, which is shown in figure 2.4(b). $1/\tau$ is the characteristic frequency of the relaxation model, which can be defined as the resonant angular frequency ($\omega_0 = 1/\tau$). In the frequency domain, the electric dipole moment can be used to compute the susceptibility using the relation: $P(\omega) = \varepsilon_0 \chi(\omega)E$. If the condition $\varepsilon_r(\omega = 0) = \varepsilon_s$ is satisfied, the relative permittivity can be expressed as equation (2.4c), where ε_∞ is the relative permittivity at an infinite frequency. The real and imaginary parts of ε_r are expressed in equations (2.4d) and (2.4e), respectively. The resonant frequency of the Debye model can be obtained by taking the derivative of ε_I, as written in equation (2.4f) [1].

The solution of equation (2.4f) is that ω $(=1/\tau)$ equals ω_0, which proves that $1/\tau$ is the resonant frequency of the Debye model. At the resonant frequency, ε_R and ε_I both equal $(\varepsilon_s - \varepsilon_\infty)/2$. In high-K materials, the value of ε_s is far larger than the value of ε_∞. Therefore, the ε_R and ε_I values of high-K materials are both close to $\varepsilon_s/2$ at the resonant frequency. The full width at half maximum (FWHM) of the ε_I spectrum can be computed by solving equation (2.4e) with the condition $\varepsilon_I = (\varepsilon_s - \varepsilon_\infty)/4$. The difference between the two frequencies $(\Delta f = f_2 - f_1)$ is defined as the FWHM, which equals $\sqrt{3}/(\pi\tau)$.

$$P(t) = P_0 e^{-t/\tau} \tag{2.4a}$$

$$P(\omega) = \int_0^\infty P_0 e^{-t/\tau} e^{-i\omega t} dt = P_0 \left(\frac{-1}{1/\tau - i\omega}\right)(e^{-(1/\tau - i\omega)t})|_0^\infty = P_0 \left(\frac{\tau}{1 - i\omega\tau}\right) \tag{2.4b}$$

$$\varepsilon_r(\omega) = \varepsilon_\infty + \frac{P_0}{E\omega_0}\left(\frac{1}{1 - i\omega/\omega_0}\right) = \varepsilon_\infty + \frac{\varepsilon_s - \varepsilon_\infty}{1 - i\omega/\omega_0} \tag{2.4c}$$

$$\varepsilon_R = \frac{\varepsilon_s \omega_0^2 + \varepsilon_\infty \omega^2}{\omega_0^2 + \omega^2} \tag{2.4d}$$

$$\varepsilon_I = \frac{\omega\omega_0(\varepsilon_s - \varepsilon_\infty)}{\omega_0^2 + \omega^2} \tag{2.4e}$$

$$\frac{d\varepsilon_I}{d\omega} = \left[\frac{\omega_0(\varepsilon_s - \varepsilon_\infty)(\omega_0^2 + \omega^2) - 2\omega_0\omega^2(\varepsilon_s - \varepsilon_\infty)}{(\omega_0^2 + \omega^2)^2}\right] = 0 \tag{2.4f}$$

Figure 2.2 shows the ε_R and ε_I spectra of a Debye material when ε_s, ε_∞, and τ are 1000, 5, and 1 μs, respectively. $1/(2\pi\tau)$ is the resonant frequency, which is about

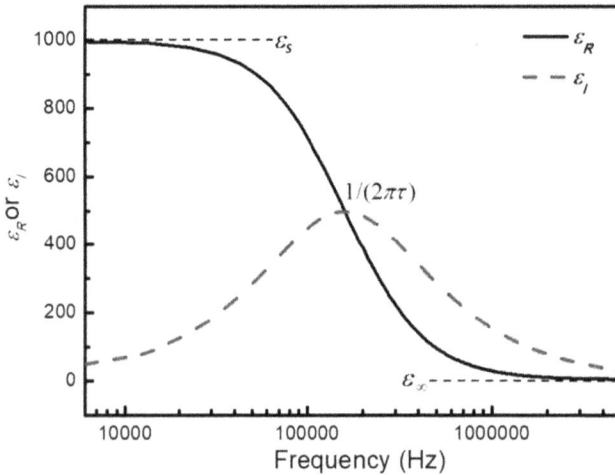

Figure 2.2. ε_R and ε_I spectra of a Debye material.

0.159 MHz. At the resonant frequency, ε_R and ε_I are both 502.5. In the low-frequency range, the ε_s value increases from 502.5 to 1000 as the frequency decreases from the resonant point to zero. In the high-frequency range, the ε_s value decreases from 502.5 to 5 as the frequency increases from the resonant point toward infinity. In addition, the FWHM of the ε_I spectrum is about 0.551 MHz because the τ value is 1 µs.

2.3 Lorentz model

The Lorentz model can be used to describe the dielectric responses of charged particles (carriers) in the presence of binding forces [2]. The motion equation of the bound carriers can be written as equation (2.5a), where F_{ext} is the external force strength, m^* is the effective mass of the bound carriers, Γ is the collision frequency, ω_0 is the characteristic frequency of the Lorentz model, a is acceleration, v is velocity, and y is the displacement of the bound carriers. When the external force (F_{ext}) is produced by the electric field of an EM wave, the motion equation of the bound carriers can be written as equation (2.5b). Conceptually, the bound carriers have simple harmonic motion because they are driven by the electric field ($E_0 e^{i(kz-\omega t)}$) of a simple harmonic electromagnetic (EM) wave, i.e. $y = y_0 e^{i(kz-\omega t)}$. Therefore, the velocity ($dy/dt = -i\omega y$) and acceleration (d^2y/dt^2) can be replaced by $-i\omega y$ and $-\omega^2 y$, respectively. The motion equation of the bound carriers can be rewritten as equation (2.5c). The displacement of the bound carriers can then be written as equation (2.5d). The induced dipole moment (p) of a bound carrier is the product of q and y, i.e. $p = qy$. We also know that the induced polarizability (α_p) of a bound carrier is the ratio of the dipole moment to the electric field. The induced polarizability can therefore be written as equation (2.5e). Using the relation ($\varepsilon_0 \chi_b = N\alpha_p$), the bound carriers' induced susceptibility can be written as equation (2.5f), where N is the density of the bound carriers. The unit of $[Nq^2/(\varepsilon_0 m^*)]^{1/2}$ is hertz, which means that this term can be viewed as a characteristic angular frequency (ω_c) of the bound carriers. Therefore, the frequency-dependent permittivity of a Lorentz material can be written as equation (2.5g), where ε_∞ is the relative permittivity at an infinite frequency.

$$F_{ext} = m^*a + m^*\Gamma v + m^*\omega_0^2 y \tag{2.5a}$$

$$qEe^{i(\omega t - kz)} = m^*[d^2y(t)/dt^2] + m^*\Gamma[dy(t)/dt] + m^*\omega_0^2 y(t) \tag{2.5b}$$

$$qE_0 = m^*(-\omega^2 y) + m^*\Gamma(-i\omega y) + m^*\omega_0^2 y \tag{2.5c}$$

$$y(\omega) = qE_0/[m^*(-\omega^2 + \omega_0^2 - i\Gamma\omega)] \tag{2.5d}$$

$$\alpha_p = q^2/[m^*(-\omega^2 + \omega_0^2 - i\Gamma\omega)] \tag{2.5e}$$

$$\chi_b = \frac{Nq^2}{\varepsilon_0 m^*[-\omega^2 + \omega_0^2 - i\Gamma\omega]} \tag{2.5f}$$

$$\varepsilon_r = \varepsilon_\infty + \frac{\omega_c^2}{[-\omega^2 + \omega_0^2 - i\Gamma\omega]} \quad (2.5g)$$

When the charged particles are electrons which are bound by an atomic nucleus, their dielectric response can be described by the Lorentz model. After light absorption, electrons in the ground state are excited to the excited state. One can imagine that the electron is connected to the atomic nucleus via a spring, as shown in figure 2.3(a). In a crystal, the mass of the atomic nucleus is far larger than that of an electron. Therefore, the electric field of an EM wave can effectively drive only the bound electrons. In a simple spring–mass model, the resonant angular frequency (ω_0) is related to the effective mass of the electron (m_e^*) and the spring constant (K). When the angular frequency of the EM wave equals the resonant angular frequency ($\omega_0 = \sqrt{K/m_e^*}$), the electron can be excited from the energy level of the valence band maximum (E_{VBM}) to the energy level of the conduction band minimum (E_{CBM}), as shown in figure 2.3(b). The difference between E_{CBM} and E_{VBM} is defined as the energy bandgap (E_g). It should be noted that the product of the wavelength (λ) and the photon energy (E_P) equals 1240 nm × eV, which can be used to compute the corresponding wavelength value when the E_g value is known. In general, semi-conductors have several absorption peaks in a broad range from the ultraviolet (UV) light band to the visible light band. In addition, the band-to-band transitions of electrons are insensitive to crystal anisotropy. Therefore, the dielectric response of a semiconductor in the UV-to-visible wavelength range can be described by the combination of the corresponding Lorentz oscillators. The relative permittivity of the Lorentz-type electronic transitions can be written as equation (2.6), where the subscript j represents the jth oscillator.

$$\varepsilon_r = \varepsilon_\infty + \sum_{j=1}^{N} \frac{f_j \omega_c^2}{[(-\omega^2 + \omega_j^2) - i\omega\Gamma_j]} \quad (2.6)$$

When the charged particles are positive and negative atoms periodically arranged in a crystal structure, the dielectric responses can be described using the modified Lorentz model, which can be expressed as equation (2.7a), where ω_{TO} is the angular frequency of the transverse optical (TO) phonon mode, ω_{LO} is the angular frequency of the longitudinal optical (LO) phonon mode, and Γ is the average collision frequency. Considering the Lyddane–Sachs–Teller relation, the relationship between

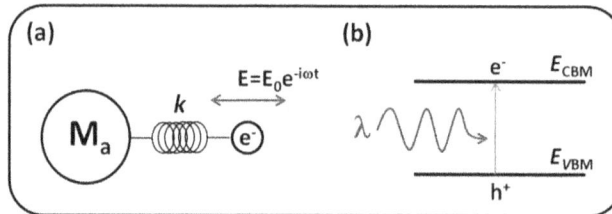

Figure 2.3. (a) A spring model. (b) Light absorption and energy diagram.

ω_{LO} and ω_{TO} is given in equation (2.7b). The relative permittivity of the modified Lorentz model can now be expressed as shown in equation (2.7c). A comparison of equations (2.5g) and (2.7c) shows that TO-LO coupling leads to modifications in the effective mass and collision rate. The effective mass of the charged particles (m_c^*) in the modified Lorentz model can be computed using equation (2.7d), where N_c is the density of charged particles. In addition, the collision frequency is doubled in the modified Lorentz model. It is notable that the charged particles of the modified Lorentz model oscillate in the transverse direction. However, the charged particles in the crystal structure can oscillate in the transverse and longitudinal directions. In other words, the modified Lorentz model is used to reduce two-dimensional oscillation to one-dimensional oscillation, which explains the doubled collision frequency (2Γ). The doubled collision frequency results in a doubled peak width in the absorption spectrum, which indicates that the TO and LO phonon modes are merged. On the other hand, the ratio of ω_{LO} to ω_{TO} can be used to compute the ratio of ε_s to ε_∞.

$$\varepsilon_r = \varepsilon_\infty + \varepsilon_\infty \left[\frac{\omega_{LO}^2 - \omega_{TO}^2}{(-\omega^2 + \omega_{TO}^2) - i(2\Gamma)\omega} \right] \tag{2.7a}$$

$$\frac{\omega_{LO}}{\omega_{TO}} = \left[\frac{\varepsilon_s}{\varepsilon_\infty} \right]^{1/2} \tag{2.7b}$$

$$\varepsilon_r = \varepsilon_\infty + \frac{(\varepsilon_s - \varepsilon_\infty)\omega_{TO}^2}{(-\omega^2 + \omega_{TO}^2) - i(2\Gamma)\omega} \tag{2.7c}$$

$$\frac{N_c e^2}{\varepsilon_0 m_c^*} = (\varepsilon_s - \varepsilon_\infty)\omega_{TO}^2 \tag{2.7d}$$

2.4 Drude model

The Drude model can be used to describe the dielectric responses of free carriers in a crystal [3]. When a free carrier is driven by the electric field of an EM wave, the equation for carrier motion can be written as equation (2.8a), where m^* is the effective mass of the carrier, qE is the electric force produced by the EM wave, a is the acceleration of the carrier, v is the drift velocity of the carrier, and τ is related to the relaxation time of the carrier. The carrier–carrier interaction is ignored due to the assumption of collective carrier oscillation in the metal crystal. Carrier relaxation mainly results from charged carrier–phonon collisions. If we replace a with dv/dt, the differential equation for v can be written as equation (2.8b). Ohm's law can be used to replace the carrier velocity (v) with an electric field (E), as written in equation (2.8c), where J is the current density and σ is the conductivity. With simple harmonic excitation, i.e. $E = E_0 e^{i(kz-\omega t)}$, we can obtain $dE/dt = -i\omega E$. By inserting $v = \sigma E/(Nq)$ and $dE/dt = -i\omega E$ into equation (2.8b), we can write the frequency-dependent conductivity of a Drude material as equation (2.8d). The differentiation of the electric constitutive relation can be written as shown in equation (2.8e). By inserting $\vec{D} = \varepsilon_0\varepsilon_r\vec{E}$, $\partial\vec{E}/\partial t = -i\omega\vec{E}$, and $\vec{P} = \varepsilon_0\chi\vec{E}$ into equation (2.8e), we can rewrite it as equation (2.8f). By replacing $\varepsilon_0\chi(-i\omega\vec{E})$ with $\sigma\vec{E}$, we can rewrite it as equation (2.8g). By combining equations (2.8d) and (2.8f), we can write the relative permittivity of a Drude material as equation (2.8h), where $Nq^2/(m^*\varepsilon_0)$ is defined as the square of the plasma frequency (ω_p) of the free carriers of a material. It should be noted that the plasma is a collective oscillation of carriers. Therefore, weak electron–electron collisions can be ignored. In fact, the Drude model has been accurately used to describe the optical responses of various metals, which confirms that electron–electron collisions are extremely weak in Au, Ag, Al, and Cu crystals.

$$qE = m^*a + \frac{m^*}{\tau}v \tag{2.8a}$$

$$qE = m^*(\mathrm{d}v/\mathrm{d}t) + (m^*v)/\tau \tag{2.8b}$$

$$J = Nqv = \sigma E \tag{2.8c}$$

$$\sigma(\omega) = \frac{Nq^2\tau}{m^*(1 - i\omega\tau)} \tag{2.8d}$$

$$\frac{\partial\vec{D}}{\partial t} = \frac{\partial(\varepsilon_0\vec{E} + \vec{P})}{\partial t} \tag{2.8e}$$

$$\varepsilon_0\varepsilon_r(-i\omega\vec{E}) = \varepsilon_0(-i\omega\vec{E}) + \varepsilon_0\chi(-i\omega\vec{E}) \tag{2.8f}$$

$$\varepsilon_0\varepsilon_r(-i\omega\vec{E}) = \varepsilon_0(-i\omega\vec{E}) + \sigma\vec{E} \tag{2.8g}$$

$$\varepsilon_r(\omega) = 1 - \left(\frac{Nq^2}{m^*\varepsilon_0}\right)/(\omega^2 + i\omega/\tau) = 1 - \omega_p^2/(\omega^2 + i\omega/\tau) \tag{2.8h}$$

We now discuss the dielectric responses of a lossless Drude medium at different EM wave frequencies. In a lossless Drude model, the formula for relative permittivity can be written as $\varepsilon_r(\omega) = 1 - (\omega_p^2/\omega^2)$. When the angular frequency ω of the EM wave is lower than the plasma frequency ω_p, the relative permittivity has a negative value, thereby changing the direction of the electric field while keeping the direction of the magnetic field in the lossless Drude medium. The constitutive relations $\vec{D} = \varepsilon_r\varepsilon_0\vec{E}$ and $\vec{B} = \mu_r\mu_0\vec{H}$ can be used to understand the change in the directions of the electric field ($\vec{E}$) and the magnetic field ($\vec{B}$). Here, $\vec{D}$ and $\vec{B}$ are continuous vector fields across the air/Drude medium interface, which results in the two relations: $\varepsilon_r^{\mathrm{air}}\varepsilon_o\vec{E}^{\mathrm{air}} = \varepsilon_r^{\mathrm{Drude}}\varepsilon_o\vec{E}^{\mathrm{Drude}}$ and $\mu_r^{\mathrm{air}}\mu_o\vec{H}^{\mathrm{air}} = \mu_r^{\mathrm{Drude}}\mu_o\vec{H}^{\mathrm{Drude}}$, where μ_r^{air} and μ_r^{Drude} are both unity. In addition, the relation $\vec{k} = C\vec{E}\times\vec{H}$ can be used to determine the propagation direction of the radiated EM wave from the driven free carriers in a lossless Drude medium, where C is a positive real number. In the air and lossless Drude medium regions, the propagation vectors are $\vec{k}^{\text{air}} = C\vec{E}^{\text{air}} \times \vec{H}^{\text{air}}$ and $\vec{k}^{\text{Drude}} = C(\vec{E}^{\text{air}} \times \vec{H}^{\text{air}})/\varepsilon_r^{\text{Drude}}$, respectively. Therefore, the relation $\vec{k}^{\text{Drude}} = -\vec{k}^{\text{air}}/\left|\varepsilon_r^{\text{Drude}}\right|$ is obtained, which explains the high reflection at the interface between the air and the lossless Drude medium when the angular frequency of the EM wave is lower than the plasma frequency. When the angular frequency of the EM wave approaches infinity, the relative permittivity approaches unity, which indicates that free carriers do not effectively interact with high-frequency EM waves. In other words, high-frequency EM waves can propagate in a lossless Drude medium.

In the near-infrared wavelength range from 800 to 3000 nm, the relative permittivity of transparent conductive materials (TCMs) can be modeled using the Drude model. Indium tin oxide (ITO), aluminum-doped zinc oxide (AZO), and poly(3,4-ethylenedioxythiophene) polystyrene sulfonate (PEDOT:PSS) are widely used as TCMs in optoelectronic devices, such as solar cells, light-emitting diodes, photodetectors, and laser diodes. In ITO and AZO films, the free carriers are free electrons. In PEDOT:PSS films, the free carriers are positive bipolarons [4]. The carrier concentration (N), effective mass (m^*), and relaxation time (τ) values of ITO, AZO, and PEDOT:PSS thin films are listed in table 2.1. The calculated plasma frequency values of ITO, AZO, and PEDOT:PSS thin films are 3.03×10^{15} Hz, 2.79×10^{15} Hz, and 2.95×10^{15} Hz, respectively. It should be noted that their plasma frequency (wavelength) values are about 3.03×10^{15} Hz (700 nm), which explains why ITO, AZO, and PEDOT:PSS films can be used as TCMs in the visible wavelength range. When the EM wave frequency is zero, equation (2.8d) is reduced to equation (2.9a). The direct current (DC) electrical conductivity (σ_{DC}) of an ITO, AZO, or PEDOT:PSS thin film can be obtained if N, m^*, and τ are known. The calculated DC conductivity values for ITO, AZO, and PEDOT:PSS thin films are 6446, 4053, and 1284 S cm^{-1}, respectively. It should be noted that the carrier type, carrier concentration, and carrier mobility of TCMs can be determined using the Hall effect. The carrier mobility is called Hall mobility; this can be used to determine the relaxation time of carriers, which is written in equation (2.9b). However, the effective mass of carriers in a material is usually determined using the cyclotron resonance effect. The cyclotron resonance frequency is inversely proportional to the effective mass, which is written as equation (2.9c), where B is an external magnetic field. When a fixed magnetic field is applied to the material, the path of a drift carrier

Table 2.1. Carrier properties of three transparent conductive materials (TCMs).

TCM	Type of carrier	N (1 cm^{-3})	m^* (m_e)	τ (fs)	References
ITO	Electron	1.01×10^{21}	0.35	7.95	[5]
AZO	Electron	6.88×10^{20}	0.28	5.87	[6]
PEDOT:PSS	Hole	8.23×10^{20}	0.30	1.66	[7]

describes a circular motion with a radius known as the cyclotron radius. The cyclotron radius can be evaluated using equation (2.9d), where v is the drift velocity of the carrier, which can be obtained using equation (2.8c). It should be noted that the cyclotron radius is about 53.4 pm when m^*, σ_{DC}, E, N, e, and B are 0.3 m_e, 5000 S cm^{-1}, 1 V cm^{-1}, 1×10^{21} cm^{-3}, 1.6×10^{-19} C, and 1 T, respectively, which shows that the effective mass is related to the local properties of the material. When the applied electric field (E) increases from 1 to 100 V cm^{-1}, the cyclotron radius increases from 53.4 pm to 5.34 nm, which means that the effective mass is related to material properties at the nanoscale.

The effective mass of the carriers in a material can be obtained from equation (2.8h) if the carrier concentration (N) is known. Conceptually, N depends on the local properties. Therefore, N can be accurately obtained by carrying out a Hall measurement. To confirm that the effective mass of carriers in a material is related to the local properties, the drift distance of the carriers is computed using equation (2.9e), where f is the frequency of the EM wave. The drift distance of the carriers is 15.6 fm when σ_{DC}, E, N, e, and f are 5000 S cm^{-1}, 1 V cm^{-1}, 1×10^{21} cm^{-3}, 1.6×10^{-19} C, and 10^{14} Hz, respectively, which shows that the effective mass and relaxation time are related to the local properties of the material. It should be noted that the free carrier concentration at the grain boundaries is very low. Therefore, the two relaxation times obtained from equations (2.8h) and (2.9b) can be used to distinguish the carrier dynamics in the crystals and at the grain boundaries by applying equation (2.9f), where τ_{Hall} is the relaxation time from equation (2.9b), τ_{Drude} is the relaxation time from equation (2.8h), and τ_{GB} is related to the collision process at the grain boundaries.

$$\sigma_{DC} = \frac{Ne^2\tau}{m^*} \tag{2.9a}$$

$$\mu = e\tau/m^* \tag{2.9b}$$

$$f_{cyclotron} = qB/(2\pi m^*) \tag{2.9c}$$

$$R_{cyclotron} = m^*v/(eB) \tag{2.9d}$$

$$d_{drift} = v/(2f) = \sigma E/(2Nef) \tag{2.9e}$$

$$1/\tau_{Hall} = 1/\tau_{Drude} + 1/\tau_{GB} \tag{2.9f}$$

In the UV-to-visible wavelength range from 200 to 700 nm, the relative permittivity of noble metals can be modeled using the Lorentz–Drude model by combining equations (2.6) and (2.8h), which can be written as equation (2.10). Equation (2.10) can be reduced to equation (2.8h) when ε_∞, f_0, and ω_0 are one, one, and zero, respectively. In other words, the dielectric response of noble metals can be described by the linear combination of the different Lorentz oscillators and free carriers (plasma). Au, Ag, Al, and Cu are widely used in optoelectronic and sensing devices owing to their relatively low ohmic losses. Values of f_j, ω_p, Γ_j, and ω_j are

Table 2.2. Values of ω_p, f_j, Γ_j, and ω_j for Au, Ag, Al, and Cu. The units of $2\pi h\Gamma$, $2\pi h\omega$, and $2\pi h\omega_p$ are eV. h is the Planck constant [8].

Metal	Metal			
	Au	Ag	Al	Cu
$2\pi h\omega_p$	9.03	9.01	14.98	10.83
f_0	0.760	0.845	0.523	0.577
$2\pi h\Gamma_0$	0.053	0.048	0.047	0.030
$2\pi h\omega_0$	0	0	0	0
f_1	0.024	0.065	0.227	0.061
$2\pi h\Gamma_1$	0.241	3.886	0.333	0.378
$2\pi h\omega_1$	0.415	0.816	0.162	0.291
f_2	0.010	0.124	0.050	0.104
$2\pi h\Gamma_2$	0.345	0.452	0.312	1.056
$2\pi h\omega_2$	0.830	4.481	1.544	2.957
f_3	0.071	0.011	0.166	0.723
$2\pi h\Gamma_3$	0.870	0.065	1.351	3.213
$2\pi h\omega_3$	2.969	8.185	1.808	5.300
f_4	0.601	0.840	0.030	0.638
$2\pi h\Gamma_4$	2.494	0.916	3.382	4.305
$2\pi h\omega_4$	4.304	9.083	3.473	11.18
f_5	4.384	5.646	None	None
$2\pi h\Gamma_5$	2.214	2.419	None	None
$2\pi h\omega_5$	13.32	20.29	None	None

listed for Au, Ag, Al, and Cu in table 2.2. If we take Au as an example, the sum of f_0, f_1, f_2, and f_3 gives a value of 0.865, which means that the electrons of the lower-energy Lorentz oscillators ($j = 1$, 2, and 3) are mainly contributed by free electrons (plasma). The f_0, f_1, f_2, and f_3 values of Au are 0.760, 0.024, 0.010, and 0.071, respectively. It should be noted that the f_0, f_1, and f_2 values of Cu are 0.577, 0.061, and 0.104, respectively, which means that it is possible to decrease the plasma frequency of Cu using near-infrared optical excitation when the two low-energy Lorentz oscillators (0.291 and 2.957 eV) are physically present. In the Lorentz–Drude model of Cu, the resonant energy of the 2nd Lorentz oscillator ($j = 2$) is 2.957 eV; therefore, Cu absorbs short-wavelength light in the visible wavelength range, which explains the reddish-yellow color of Cu. In other words, the 2nd Lorentz oscillator (bound electrons) physically and mathematically manifests in Cu metal, reducing the free carrier density and thereby decreasing the plasma frequency. In regard to Cu, the f_3 and f_4 values are 0.723 and 0.638, respectively. The sum of f_3 and f_4 is larger than that of f_0, f_1, and f_2, which indicates that the electrons of the higher-energy Lorentz oscillators ($j = 3$ and 4) do not originate as free electrons. In other words, light absorption at the two higher-photon energy values (5.30 and 11.18 eV) can increase the free carrier density of Cu and thereby increase its plasma angular frequency. However, the two higher-photon energy values are larger than the work function of Cu, which can result in the light-induced oxidation of Cu in an oxygen environment. The work function of Cu is about 5.0 eV when the crystal plane is (111). The above discussions can also be used to understand the relation between free carriers and bound carriers in Ag and Al metals.

$$\varepsilon_r = \varepsilon_\infty + \sum_{j=0}^{N} \frac{f_j \omega_p^2}{[(-\omega^2 + \omega_j^2) - i\omega\Gamma_j]} \tag{2.10}$$

2.5 Transfer matrix method

As mentioned in chapter 1, wave plates can be described by Jones matrices when an EM wave propagates in a fixed direction. Such matrices are widely used to understand the polarization manipulation of a Gaussian beam (the definition of a Gaussian beam will be given in chapter 3). A Gaussian beam can be viewed as a plane wave when the divergence angle is small. When a Gaussian beam propagates in a material, the propagation direction and refractive index can be used to describe wave propagation; these can be expressed in the form of a matrix. The matrix of a film can be written as equation (2.11a), where Δ and Y denote the optical phase and the optical admittance, respectively. Δ is proportional to the propagation constant and path length, which is written as equation (2.11b), where n is the refractive index, λ_0 is the free-space wavelength, and L is the path length. Y is related to the material properties and is given by equation (2.11c), where ε, μ, and η_0 are the permittivity, permeability, and free-space wave impedance, respectively.

$$M = \begin{bmatrix} \cos(\Delta) & i\,\sin(\Delta)/Y \\ iY\,\sin(\Delta) & \cos(\Delta) \end{bmatrix} \tag{2.11a}$$

$$\Delta = kL = n(2\pi/\lambda_0)L \tag{2.11b}$$

$$Y = \sqrt{\varepsilon/\mu} = (\sqrt{\varepsilon_r/\mu_r})/\eta_0 \tag{2.11c}$$

If we consider a uniform thin film in a tri-layered structure (see figure 2.4), the relation between the EM waves in region 0 and region 2 can be written as equation (2.12a). In a non-magnetized material ($\mu_r = 1$), Y equals n. The EM wave is normally incident from region 0, which results in a reflectance (R) and a

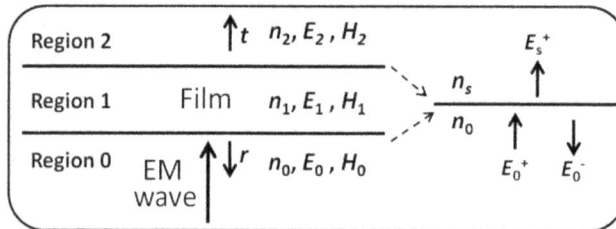

Figure 2.4. A tri-layered structure and the effective model.

transmittance (T). Regions 1 and 2 can be viewed as a substrate. Therefore, the tri-layered structure can be reduced to an air/substrate structure. The transmission coefficient (t) and reflectance coefficient (r) of a non-magnetized thin film are defined as $E_b/E_{0a}^+ = 2n_0 E_2/(n_0 E_0 + H_0)$ and $E_b/E_{0a}^+ = (n_0 E_0 - H_0)/(n_0 E_0 + H_0)$, respectively. It should be noted that E_0/H_0 can be defined as the effective refractive index of the non-magnetized substrate (n_s). Using equation (2.12a), the relations between E_2, H_2, E_0, and H_0 can be obtained, where Δ_1 and n_1 are the optical phase and refractive index in region 1. Then, t and r can be rewritten as equations (2.12b) and (2.12c), respectively, where the E_2/E_0 and n_s values must be determined from equation (2.12a). $T = [\mathrm{Re}(n_s)/n_0]|t|^2$ and $R = |r|^2$ can be used to determine the transmittance and reflectance values [9]. The correctness of equations (2.12b) and (2.12c) can be tested if the thickness (L_1) of the non-magnetized film is zero. When L_1 equals zero, t and r reduce to $2n_0/(n_0 + n_2)$ and $(n_0 - n_2)/(n_0 + n_2)$, respectively. T and R equal 96% and 4% when n_0 and n_2 are 1.0 and 1.5, respectively, which shows that a glass plate ($n = 1.5$) has a high transmittance of 92% and a low reflectance of 8% in the visible wavelength range.

$$\begin{bmatrix} E_0 \\ H_0 \end{bmatrix} = \begin{bmatrix} \cos(\Delta_1) & i\,\sin(\Delta_1)/n_1 \\ i n_1 \sin(\Delta_1) & \cos(\Delta_1) \end{bmatrix} \begin{bmatrix} E_2 \\ H_2 \end{bmatrix} \tag{2.12a}$$

$$t = \frac{2n_0(E_2/E_0)}{n_0 + n_s} \tag{2.12b}$$

$$r = \frac{n_0 - n_s}{n_0 + n_s} \tag{2.12c}$$

If we consider a metal-coated glass substrate, the transmittance and reflectance spectra can be predicted using equations (2.12b), (2.12c), and (2.10). The relative permittivity spectrum of Cu and the transmittance and reflectance spectra of a Cu/glass substrate are plotted in figure 2.5. The thickness of the Cu film is 10 nm. The refractive index of the glass substrate is assumed to have the same value of 1.5 over the visible wavelength range. Equations (2.12b) and (2.12c) show that the optical

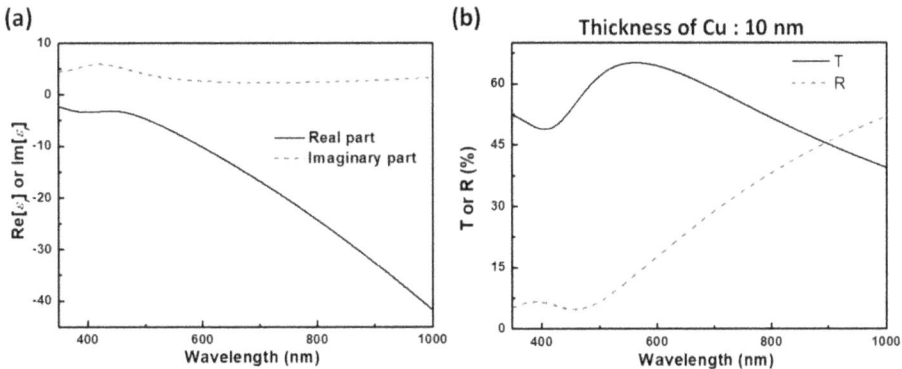

Figure 2.5. (a) Wavelength-dependent relative permittivity of Cu. (b) Transmittance and reflectance spectra of a Cu/glass substrate.

responses of the Cu/glass substrate are strongly related to the optical phase (Δ) and refractive index (n) of the Cu film. In other words, the thickness and refractive index of an unknown film coated on a glass substrate can be determined by analyzing the measured transmittance or reflectance spectrum with a curve fitting method. In the curve fitting process, the thickness and refractive index can be assigned as variables to which the experimental data points are fitted.

If we consider a multilayered structure (see figure 2.6), the relation between the EM waves in region 0 and region $m + 1$ can be written as equation (2.13a). The transmission coefficient (t) and reflectance coefficient (r) of the multilayered structure are defined as $2n_0(E_{m+1}/E_0)/(n_0 + n_s)$ and $(n_0 - n_s)/(n_0 + n_s)$, where $E_{m + 1}$ is the electric field in region $m + 1$, E_0 is the electric field in region 0, and n_s is the effective refractive index of the substrate. Then, t and r can be rewritten as equations (2.13b) and (2.13c), respectively, where n_0, n_{m+1}, E_0, and H_0 are the refractive index in region 0, the refractive index in region $m + 1$, the electric field in region 0, and the magnetic field in region 0, respectively. $T = [\mathrm{Re}(n_s)/n_0]|t|^2$ and $R = |r|^2$ can be used to determine the transmittance and reflectance values, respectively.

$$\begin{bmatrix} E_0 \\ H_0 \end{bmatrix} = M_1 \cdots M_{m-2} M_{m-1} M_m \begin{bmatrix} E_{m+1} \\ H_{m+1} \end{bmatrix} = M_{\mathrm{eff}} \begin{bmatrix} E_{m+1} \\ H_{m+1} \end{bmatrix} = \begin{bmatrix} m_{11}^{\mathrm{eff}} & m_{12}^{\mathrm{eff}} \\ m_{21}^{\mathrm{eff}} & m_{22}^{\mathrm{eff}} \end{bmatrix} \begin{bmatrix} E_{m+1} \\ H_{m+1} \end{bmatrix} \quad (2.13a)$$

$$t = \frac{2n_0(E_{m+1}/E_0)}{n_0 + (H_{m+1}/E_{m+1})} \quad (2.13b)$$

$$r = \frac{n_0 - (H_{m+1}/E_{m+1})}{n_0 + (H_{m+1}/E_{m+1})} \quad (2.13c)$$

Equations (2.13a), (2.13b), and (2.13c) can be used to design optical filters. Let us consider a 42-layer structure, in which the bottom layer is free space, the top layer is a glass substrate, and the middle 40 layers are formed of 20 pairs of a low-index thin

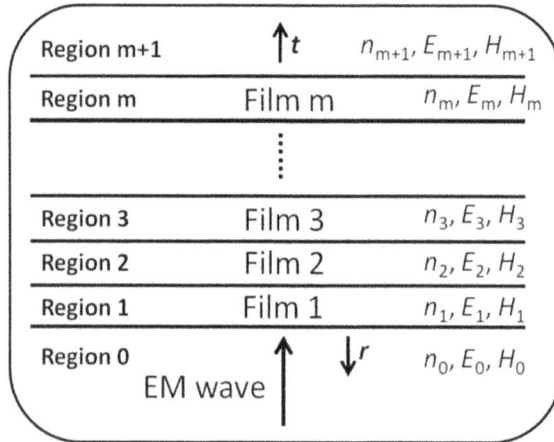

Figure 2.6. A multilayered structure.

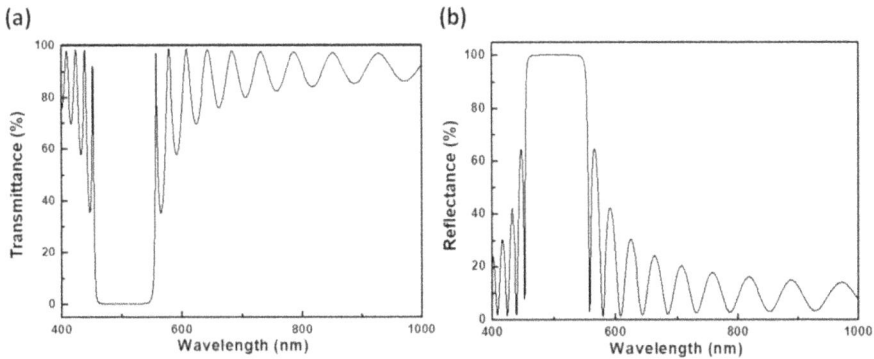

Figure 2.7. (a) Transmittance and (b) reflectance spectra of a 42-layer structure.

film and a high-index film. The transmittance and reflectance spectra of the 42-layer structure are plotted in figure 2.7 for refractive index and thickness values of the low-index (high-index) film of 1.5 (2.0) and 83.3 nm (62.5 nm), respectively. It should be noted that the thickness of the films equals $\lambda_0/(4n)$, where λ_0 is 500 nm. The 12-layer structure is also a Bragg reflector, which has a high reflectance at the design wavelength ($\lambda_0 = 500$ nm). In other words, the middle 10 layers form a forbidden band centered at a wavelength of 500 nm, which can be used as a band-stop filter.

Bibliography

[1] Mantas P Q 1999 Dielectric response of materials: extension to the Debye modle *J. Eur. Ceram. Soc.* **19** 2079–86

[2] Oughstun K E and Cartwright N A 2003 On the Lorentz–Lorenz formula and the Lorentz model of dielectric dispersion *Opt. Express* **11** 1541–6

[3] Drude P 1900 Zur Elektronentheorie der Metalle; II. Teil. Galvanomagnetische und thermomagnetische effects *Ann. Phys.* **308** 369–402

[4] Camacho-Guardian A, Ardila L A P, Pohl T and Bruun G M 2018 Bipolarons in a Bose–Einstein condensate *Phys. Rev. Lett.* **121** 013401

[5] Qiao Z, Latz R and Mergel D 2004 Thickness dependence of In_2O_3:Sn film growth *Thin Solid Films* **466** 250–8

[6] Nomoto J, Makino H and Yamamoto T 2015 Carrier mobility of highly transparent condeuctive Al-doped ZnO polycrystalline films deposited by radio-frequency, direct-current, and radio-frequency-superimposed direct-current magnetron sputtering: grain boundary effect and scattering in the grain bulk *J. Appl. Phys.* **117** 045304

[7] Chang S H, Chiang C-H, Kao F-S, Tien C-L and Wu C-G 2014 Unraveling the enhanced electrical conductivity of PEDOT:PSS thin films for ITO-free organic photovoltaics *IEEE Photonics J.* **6** 8400307

[8] Rakic A D, Djurisic A B, Elazar J M and Majewski M L 1998 Optical properties of metallic films for vertical-cavity optoelectronic devices *Appl. Opt.* **37** 5271–83

[9] Mackay T G and Lakhtakia A 2020 *The Transfer-Matrix Method in Electromagnetics and Optics* (Cham: Springer)

IOP Publishing

Light–Material Interactions and Applications in Optoelectronic Devices

Anjali Chandel and Sheng Hsiung Chang

Chapter 3

Laser beam propagation, properties, and manipulation

In this chapter, the propagation characteristics of a Gaussian beam are analyzed via Fourier transform from the spatial distribution to the momentum distribution. We show that it is possible to deduce the diffraction angle and Rayleigh range of a Gaussian beam using simple equations. In addition, lenses, beam expanders, objectives, or plasmonic lenses can be used to manipulate a Gaussian beam by functioning as collimators, focusing devices, or optical couplers. The Abbe number is an important metric that is used to describe material dispersion. Snell's law and the coupling mode equation are used to compute the coupling efficiency of a focused Gaussian beam to a single-mode dielectric waveguide. In addition, the design concept of a plasmonic lens is illustrated in order to explain the working mechanism of a planar lens. The dispersion relations of a planar surface plasmon wave and a metal-dielectric-metal plasmonic waveguide are described in the last section.

3.1 Gaussian beams

In chapter 1, we described the propagation, properties, and manipulation of an EM plane wave. However, the light wave emitted by an optical cavity-based laser is usually a Gaussian beam. At the exit port of a high-quality laser, the wave fronts of a Gaussian beam are nearly flat and parallel, so that the beam can be viewed as a plane wave. Diffraction is an intrinsic property of EM waves that has a divergent effect on wave propagation. The Fourier transform of a two-dimensional (2D) Gaussian profile can be written as equation (3.1a) in cylindrical coordinates, where $2a$ is defined as the diameter of the Gaussian beam. After solving the integration, $\tilde{E}(k)$ can be written as shown in equation (3.1b). The k-dependent intensity distribution can be obtained using the relation: $I(k) = \tilde{E}(k)\tilde{E}^*(k)$, which is expressed in equation (3.1c). It should be noted that k is related to the angle, which

doi:10.1088/978-0-7503-6099-9ch3

is expressed in equation (3.1d). By combining equations (3.1c) and (3.1d), the angle-dependent intensity distribution can be written as equation (3.1e). When the beam diameter ($2a$) is far larger than the wavelength (λ), $I(\theta)$ can be rewritten as equation (3.1f). It should be noted that the intensity distributions in real space and k space are both 2D Gaussian functions. The intensity distribution in k space is the diffraction result, which can be used to evaluate the self-diffraction characteristics of a Gaussian laser beam. The self-diffraction angle is defined as $\theta = \lambda/(\pi a)$ when $I(\theta)$ equals $e^{-2}a^2/(4\pi)$.

$$\tilde{E}(k) = \left(\frac{1}{2\pi}\right)^2 \int_{-\infty}^{\infty}\left[\int_{-\pi}^{\pi} e^{-r^2/a^2} \times e^{-ikr} r d\phi\right] dr \tag{3.1a}$$

$$\tilde{E}(k) = \frac{1}{2\pi}\left[\left(\frac{-i}{k}\right) - a\sqrt{\pi}\, e^{-\frac{1}{4}k^2a^2}\right] \tag{3.1b}$$

$$I(k) = \frac{1}{4\pi^2}\left(\frac{1}{k^2} + a^2\pi e^{-\frac{1}{2}k^2a^2}\right) \tag{3.1c}$$

$$k = \frac{2\pi}{\lambda}\sin\theta \tag{3.1d}$$

$$I(\theta) = \frac{1}{4\pi^2}\left(\frac{\lambda^2}{4\pi^2\sin^2\theta} + a^2\pi e^{-\frac{2\pi^2}{\lambda^2}a^2\sin^2\theta}\right) \tag{3.1e}$$

$$I(\theta) \sim \frac{a^2}{4\pi}e^{-\frac{2\pi^2}{\lambda^2}a^2\sin^2\theta} \tag{3.1f}$$

The propagation path of an emitted Gaussian laser beam is plotted in figure 3.1 (a). At the initial position ($z = 0$), the beam diameter is $2a$. Curves A and B are lines tracing the laser beam waist at two fixed ϕ angles (π and $-\pi$). These show that the beam size of the Gaussian laser beam increases with propagation distance, owing to the intrinsic wave-diffraction property. The radius of a Gaussian laser beam can be approximately predicted via the geometric relation expressed as $w(z) = a + z\tan\theta$. In other words, the lateral profile of a Gaussian laser beam in the x–y (r–ϕ) plane is a function of z. The electric field of a Gaussian laser beam can be written as equation (3.2a), where $u(x,y,z)$ is the lateral profile of the laser beam. At $z = 0$, the lateral profile can be expressed in Cartesian coordinates as equation (3.2b). The self-diffraction effect means that the propagation of a Gaussian laser beam in the x- and y-directions must be included in the z-dependent lateral profile. Therefore, the lateral profile $u(x,y,z)$ can be written as equation (3.2c), where $R(z)$ is the radius of curvature and $w(z)$ is the beam radius. It should be noted that $A(z)$ is a function of z, which is used to satisfy energy conservation. In equation (3.2c), the complex term can be defined as equation (3.2d). To solve $R(z)$ and $w(z)$, $q(z_0)$ is defined as q_0. Therefore, $q(z)$ can be expressed as equation (3.2e). At $z = 0$, the radius of curvature is infinite, which results in the relation: $(q_0 - z_0) = i\pi a^2/\lambda$. Therefore, $q(z)$ can

(a)

(b) z = 0

(c) z = z_L

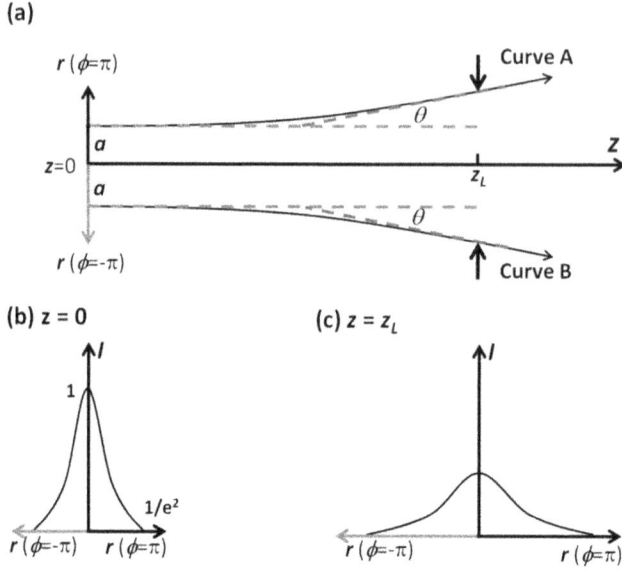

Figure 3.1. (a) Propagation path of an emitted Gaussian laser beam. (b) Gaussian beam profile at $z = 0$. (c) Gaussian profile at $z = z_L$.

rewritten as equation (3.2f), where z_R is defined as the Rayleigh range, which is proportional to the square of the initial radius. In other words, a Gaussian beam can be viewed as a plane wave when its initial radius is infinite. By combining equations (3.2d) and (3.2f), $R(z)$ and $w(z)$ can be written as equations (3.2g) and (3.2h), respectively. When z is zero, the R and w values equal infinity and a, respectively, which confirms the correctness of equations (3.2g) and (3.2h).

$$E(x, y, z) = u(x, y, z)e^{i(kz - \omega t)} \tag{3.2a}$$

$$u(x, y, 0) = \exp\left(-\frac{x^2 + y^2}{a^2}\right) \tag{3.2b}$$

$$u(x, y, z) = A(z)\exp\left[-ik\left(\frac{x^2 + y^2}{2}\right)\left(\frac{1}{R(z)} - i\frac{\lambda}{\pi w^2(z)}\right)\right] \tag{3.2c}$$

$$\frac{1}{q(z)} = \frac{1}{R(z)} - i\frac{\lambda}{\pi w^2(z)} \tag{3.2d}$$

$$q(z) = q_0 + (z - z_0) = (q_0 - z_0) + z \tag{3.2e}$$

$$q(z) = i\left(\frac{\pi a^2}{\lambda}\right) + z = iz_R + z \tag{3.2f}$$

$$R(z) = \frac{z^2 + z_R^2}{z} \tag{3.2g}$$

$$w(z) = a\sqrt{1 + \frac{z^2}{z_R^2}} \tag{3.2h}$$

Figure 3.2 displays the z-dependent radius values of a Gaussian laser beam, where the initial radius (a) and wavelength (λ) of the Gaussian laser beam are 1 mm and 500 nom, respectively. When the approximation (black line with squares) is used, overestimation of the radius mainly occurs within the Rayleigh range ($z < z_R$). The Rayleigh length (z_R) is 2π m, which is obtained from equation (3.2f). It should be noted that the divergence rate of a Gaussian laser beam increases with propagation distance (z) within the Rayleigh range. In other words, to effectively manipulate a Gaussian laser beam using miniaturized optical components, the total optical path length of the system should be shorter than the beam's Rayleigh range.

The cross-sectional area of a Gaussian laser beam can be defined as $\pi w^2(z)$, which can be written as equation (3.3a). To satisfy energy conservation, the power must be fixed. Therefore, the product of $|A(z)|^2$ and $Area(z)$ is a constant, which can be written as equation (3.3b), where C is a constant. When C equals $(\pi a^2)/z_R^2$, $|A(z)|^2$ equals $1/(z_R^2 + z^2)$. Therefore, $A(z)$ can be written as equation (3.3c). By combining equations (3.2c), (3.2d), and (3.3c), the lateral profile of a Gaussian laser beam can be rewritten in a concise formula as equation (3.3d). When the initial radius a is infinite, $1/q(z)$ equals zero, thereby resulting in a product of zero and $e^{-\infty}$ in the lateral profile $u(x,y,z)$. When the value of a approaches infinity, the value of $1/q(z)$ approaches zero, which indicates that an infinite number of photons spreads over an infinite area. When the value of a approaches infinity, the distribution function

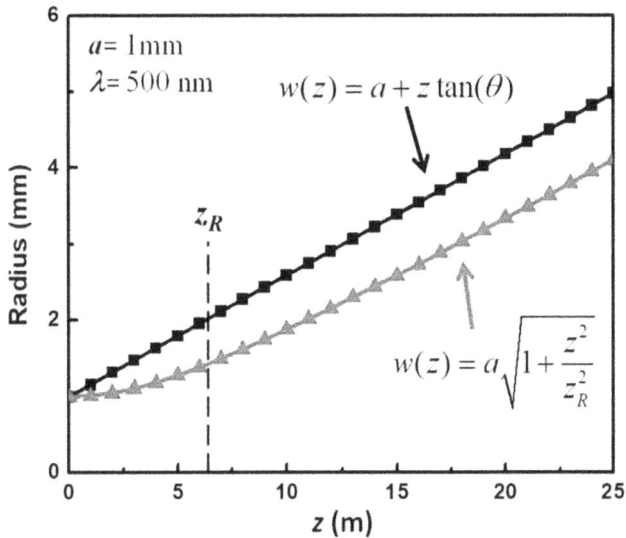

Figure 3.2. The z-dependent radius values of a Gaussian laser beam calculated using different equations.

approaches unity ($e^{-\infty}$), which indicates that the photons are uniformly distributed in the lateral (x–y) plane. In other words, the lateral profile of a Gaussian beam becomes an infinite and flat wavefront when the value of a approaches zero.

$$\text{Area}(z) = \pi w(z) = \frac{\pi a^2}{z_R^2}(z_R^2 + z^2) \tag{3.3a}$$

$$C = |A(z)|^2 \frac{\pi a^2}{z_R^2}(z_R^2 + z^2) \tag{3.3b}$$

$$A(z) = 1/(iz_R + z) \tag{3.3c}$$

$$u(x, y, z) = \frac{1}{q(z)} \exp\left[-ik\left(\frac{x^2 + y^2}{2q(z)}\right)\right] \tag{3.3d}$$

A collimated Gaussian laser beam has a self-diffraction angle in a uniform and isotropic medium, which is an intrinsic property of EM waves. When a collimated Gaussian laser beam propagates along the optical axis of a convex lens, the beam size is decreased to a diffraction-limited diameter at the focal plane, as shown in the inset of figure 3.3. The curves A and B trace the lines of the laser beam waist; they show that the Gaussian laser beam can be focused on the focal plane. In a short-focal-length lens, the distance between the lens and the focal plane (F_d) is significantly larger than the focal length (F_g), which is determined by a geometric relation. If θ_f and z_R denote the focusing angle and Rayleigh range, respectively, the radius of the Gaussian laser beam at the focal plane can be computed using equation (3.2h) when z equals the diffraction-limited focal length (F_d). The smallest radius of

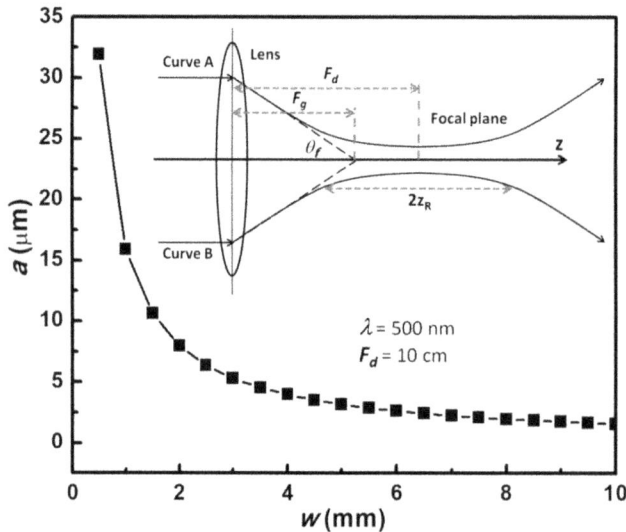

Figure 3.3. The a–w curve. Here, a and w are the radial values of the focused and collimated Gaussian beams, respectively.

the focused Gaussian beam (a) is related to the radius of the collimated Gaussian beam (w), which can be expressed as equation (3.4a). The a–w curve is plotted in figure 3.3 for F_d and λ values of 10 cm and 500 nm, respectively. The value of a decreases from about 32 μm to about 1.6 μm as the value of w increases from 0.5 to 10 mm, which shows that the value of a is limited by the self-diffraction effect. The spot size of a focused Gaussian laser beam at the focal plane is 3.2 μm when the diameter of the collimated Gaussian laser beam and the focal length of the lens are 20 mm and 10 cm, respectively, which explains the larger physical sizes of long-working-distance objectives. It should be noted that the discussion above is based on the paraxial approximation, which leads to the concept of a clear aperture when designing an objective lens. When the diameter of the collimated Gaussian laser beam is less than the clear aperture of the lens, it is possible to predict the smallest radius a of the focused Gaussian laser beam using equation (3.4a). The numerical aperture (NA) of a lens can be defined as equation (3.4b), where n is the refractive index of the environment and w_c is the radius of the clear aperture of the lens. For a one-inch lens, the calculated NA is about 0.316 when the values of w_c and F_d are 10 and 30 mm, respectively, which shows that it is difficult to increase the NA value while using a single lens.

$$a = \sqrt{\frac{w_c^2 - \sqrt{w_c^4 - 4F_d^2 \lambda^2/\pi^2}}{2}} \tag{3.4a}$$

$$\mathrm{NA} = n \sin \theta_f = n w_c / \sqrt{w_c^2 + F_g^2} \tag{3.4b}$$

3.2 Beam expanders

Beam expanders are optical systems that increase the diameter of a laser beam. They typically consist of lenses or lens systems that expand the beam size while maintaining its collimation. Figure 3.2 shows that the radius of a Gaussian laser beam increases with the propagation distance because of the self-diffraction effect. In many optical systems, it is necessary to greatly increase or decrease the radius of a Gaussian laser beam within a limited space. Fortunately, combinations of two lenses can be used to adjust the radius of the Gaussian laser beam, as shown in figure 3.4. When two convex lenses (L_1 and L_2) are used, the distance between L_1 and L_2 must equal the sum of F_1 and F_2, where F_1 and F_2 are the focal lengths of L_1 and L_2, respectively. According to their geometric relationship, the radius of the output Gaussian laser beam can be computed using equation (3.5a), where w_1 is the radius of the input Gaussian laser beam. When F_2 is larger than F_1, the radius of the Gaussian laser beam is increased from w_1 to w_2, representing a magnification of F_2/F_1. The combination of two convex lenses is known as a Keplerian beam expander, which has a large physical length. The design concept of the Keplerian beam expander originated from the Keplerian telescope, which was invented by Johannes Kepler in the 17th century. When a concave lens (L_1) and a convex lens (L_2) are used, the distance between L_1 and L_2 must equal the sum of F_1 and F_2.

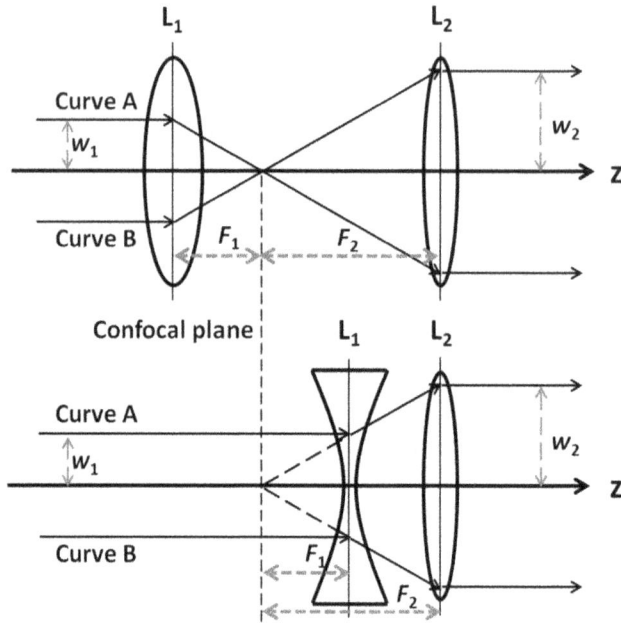

Figure 3.4. Ray-tracing lines in Keplerian (above) and Galilean (below) beam expanders.

It should be noted that F_1 has a negative value. The combination of a concave lens and a convex lens is known as a Galilean beam expander; this approach greatly decreases the length of the beam expander. The size of the Galilean beam expander is given by equation (3.5b).

When a Galilean beam expander is integrated with an optical imaging system, the image orientation remains unchanged. When a Keplerian beam expander is integrated with an optical imaging system, the image quality can be improved by the use of a small pinhole placed at the focal point. The aperture (diameter) of the pinhole must be slightly larger than the diameter of the Gaussian laser beam at the focal plane, which can be used as a spatial filter. In other words, the focal plane can be viewed as a 2D momentum space.

$$w_2 = w_1 \left(\frac{F_2}{F_1} \right) \tag{3.5a}$$

$$M = \frac{w_2}{w_1} = \left| \frac{F_2}{F_1} \right| \tag{3.5b}$$

3.3 Objective lenses

Objective lenses are primarily used in microscopy and imaging systems to form images of micrometer-sized objects. Figure 3.5 shows that the NA value can be increased by reducing the focal length when two convex lenses (L_1 and L_2) are

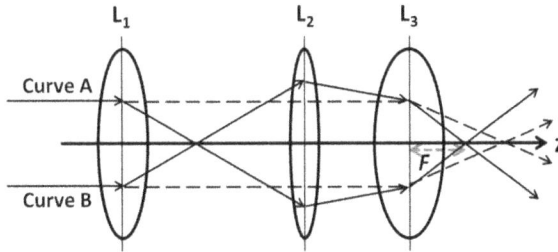

Figure 3.5. Ray-tracing lines in a system of compound lenses.

Table 3.1. Optical properties of various lens materials.

Material	Refractive index	Abbe number	Transparent wavelength range (μm)	References
Fused silica	1.462 @ 0.5 μm	67.8	0.21–3.71	[1]
N-BK7	1.521 @ 0.5 μm	64.1	0.30–2.50	[2]
N-SF11	1.803 @ 0.5 μm	25.7	0.37–2.50	[2]
MgF$_2$	1.379 @ 0.5 μm	106.3	0.14–7.50	[3]
CaF$_2$	1.436 @ 0.5 μm	95.3	0.15–12	[4]
BaF$_2$	1.478 @ 0.5 μm	81.8	0.15–15	[4]
Si	3.479 @ 1.5 μm	None	1.20–14	[4]

combined. The focal length (F) is defined as the distance between the convex lens (L_3) and the focal point. The focal point is the intersection point of curves A and B after the third convex lens. When L_1 and L_2 are removed, the ray-tracing lines A and B are replaced by the two dashed lines, which results in a longer focal length. The physical length of an objective is mainly determined by the distance between L_1 and L_3. To reduce the physical length of an objective, the first convex lens (L_1) can be replaced by a concave lens, which decreases the distance between L_1 and L_2.

Objective lenses are designed for use in optical microscopes that operate in the broad wavelength range from 200 to 5000 nm. Optical lenses are fabricated using borosilicate (N-BK7) glass, fused silica, N-SF11, CaF$_2$, MgF$_2$, Si, and BaF$_2$. The refractive indices, Abbe numbers, and transparent wavelength ranges of the above-mentioned optical lenses are listed in table 3.1. The Abbe number (V_d) can be computed using equation (3.6), where n_d, n_F, and n_C are the refractive index values at wavelengths of 587.56, 486.1, and 656.3 nm, respectively. The main emission peak of helium plasma has a wavelength of 587.56 nm. The two main emission peaks of hydrogen plasma have wavelengths of 486.1 and 656.3 nm, respectively. Early scientists used helium gas and hydrogen gas lamps as light sources in order to evaluate material dispersion. Larger Abbe numbers represent lower material dispersion in the wavelength range from 486.1 to 656.3 nm.

$$V_d = \frac{n_d - 1}{n_F - n_C} \tag{3.6}$$

In the three SiO$_2$-glass-based materials, the trend of the Abbe number values is inversely proportional to the trends of the refractive index and the edge wavelength values. At a wavelength of 0.5 μm, the refractive index value increases from 1.462 to 1.803, while the edge wavelength increases from 0.21 to 0.37 μm. It should be noted that the edge wavelength can also be viewed as the bandgap wavelength (λ_g). The bandgap energy (E_g) can be computed using the relation: $E_g = 1240/\lambda_g$. In other words, the larger bandgap results in a lower refractive index and a larger Abbe number.

The bandgap values of MgF$_2$, CaF$_2$, and BaF$_2$ are 10.8, 10.6, and 10.5 eV, which correspond to wavelengths of 0.114, 0.117, and 0.118 μm, respectively. Therefore, the edge wavelength values at 0.14, 0.15, and 0.15 cannot be viewed as the bandgap wavelength values of MgF$_2$, CaF$_2$, and BaF$_2$, respectively. The larger bandgaps of the three metal fluorides result in lower refractive indices and larger Abbe numbers. The larger Abbe numbers show that MgF$_2$, CaF$_2$, and BaF$_2$ are suitable for use as low-dispersion lens materials.

The Abbe number of Si cannot be defined because its transparent wavelength range does not cover the wavelength range from 486.1 to 656.3 nm. To evaluate the material dispersion, a modified Abbe number (U_d) can be computed in the operational wavelength range using equation (3.7a), where n_O is the refractive index at the operating wavelength, n_0 is the refractive index at zero frequency, n_H is the refractive index at 0.8 times the operating wavelength, and n_L is the refractive index at 1.2 times the operating wavelength. When Si lenses are designed for use in the wavelength range from 4 to 6 μm, the operational wavelength (λ_O) is 5 μm. Here, n_O, n_H, and n_L are the refractive index values of Si at 5, 4, and 6 μm, respectively. In the operational wavelength range from 4 to 6 μm, the n_O, n_H, and n_L values for Si are 3, 3.420, 3.4230, and 3.418, respectively. The modified Abbe number of Si is 84.0 in the operational wavelength range from 4 to 6 μm, which shows that Si is suitable for use as a low-dispersion lens material in the transparent wavelength range. The focal length of a Si plano-convex lens can be computed using equation (3.7b), where n and R are the refractive index and the radius of curvature, respectively. When R is 10 cm, the F_H, F_O, and F_L values of a Si lens at wavelengths of 4, 5, and 6 μm are 29 214, 29 234, and 29 257 μm, respectively. Equation (3.7c) can be used to evaluate the spatial dispersion of the focal length. Under the abovementioned conditions, the S_d value of a Si lens is 10.75.

$$U_d = \frac{n_O - n_0}{n_H - n_L} \tag{3.7a}$$

$$F = \frac{R}{n - 1} \tag{3.7b}$$

$$S_d = \frac{F_L - F_H}{\lambda_O} \tag{3.7c}$$

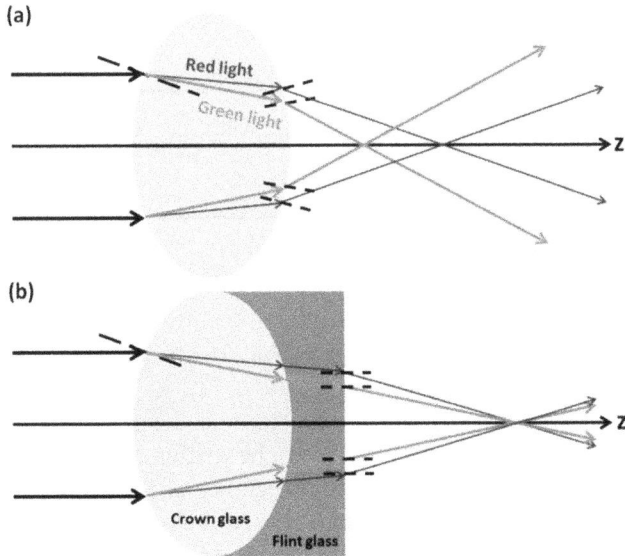

Figure 3.6. Ray-tracing lines for different wavelengths in a singlet lens and a doublet lens.

A doublet lens can be used to compensate for spatial dispersion in the focal length, as shown in figure 3.6. Red and green rays represent longer-wavelength and shorter-wavelength photons, respectively. The dashed lines represent the surface normal to the corresponding interfaces. At the air/glass interface, the transmitted angle of the red light is larger than that of the green light due to the positive dispersion relation. Therefore, the focal length of red light is longer than that of green light, as shown in figure 3.6(a). In doublet lens components, N-BK7 and N-SF11 materials are widely used as the crown and flint glasses, respectively. At the crown glass/flint glass interface, the refractive index (incident angle) of the red light is smaller (larger) than that of the green light, which results in larger and smaller transmitted angles for the red and green rays in the flint glass, respectively, thereby forming the nearly parallel red and green rays in the flint glass. At the flint glass/air interface, the transmitted angle of the red light is larger than that of the green light, thereby forming a common focal point for the green and red rays. The above discussion explains the main working mechanism of achromatic lenses.

The simplest objective lens contains three lenses; it therefore has six interfaces and thereby offers relatively lower transmittance due to reflections. At an air/glass interface, the transmittance is about 96% in the visible and near-infrared wavelength ranges. The calculated result shows that the transmittance of the three glass lenses is about 78.27%. Therefore, it is necessary to reduce the reflection loss. Fortunately, a thin film can be used to reduce reflection loss by forming destructive interference in the reflection. Conceptually, the optical path length of the thin film must be a quarter wavelength in order to create destructive interference, as shown in figure 3.7. When considering the first reflections from the a and b interfaces, the reflected electric field (E_r) can be written as equation (3.8a), where t, r, k_f, and L_f are the

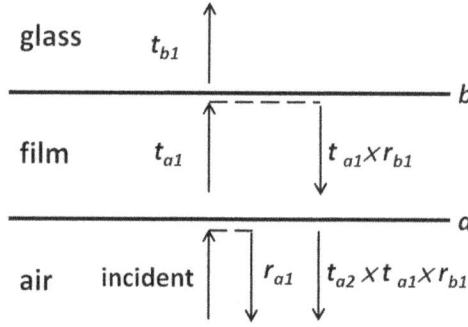

Figure 3.7. Concept of interference in a tri-layered structure.

transmissivity, reflectivity, propagation constant of the film, and thickness of the film, respectively. When $2k_f L_f$ equals π, E_r can be written as $E_0(r_{a1} - t_{a2} t_{a1} r_{b1})$, which is defined as destructive interference. The reflectance of the tri-layered structure can be computed using equations (2.12a) and (2.12c) if the refractive index and thickness of the film are known. To completely reduce the reflection loss, the required refractive index of the quarter-wavelength film can be determined by solving equation (2.12c), which is written as equation (3.8b), where n_{air} and n_{glass} are the refractive index values of air and glass, respectively. The value of n_f is about 1.225 when n_{air} and n_{glass} are 1.0 and 1.5, respectively. In other words, the thickness $(L_f = \lambda_0/(4n_f))$ and refractive index of the quarter-wavelength film are about 100.2 nm and 1.225 at $\lambda_0 = 500$ nm, respectively. Therefore, the quarter-wavelength film is also called an antireflective (AR) layer.

$$E_r = E_0 r_{a1} + E_0 t_{a2} t_{a1} r_{b1} e^{i2k_f L_f} \tag{3.8a}$$

$$n_f^2 = n_{air} \times n_{glass} \tag{3.8b}$$

3.4 Optical couplers

Optical couplers play an important role in optical communication systems, since they are used to effectively connect lasers (detectors) to optical integrated circuits (OICs). To completely couple a laser beam to an OIC, the modal profiles of the laser Gaussian beam and the waveguide must spatially overlap. To simplify this problem, the 1D modal profiles of a laser Gaussian beam and a dielectric waveguide are plotted in figure 3.8, where E_{FL} and E_{WG} are the electric field distributions of the focused laser Gaussian beam and waveguide mode in the x-direction, respectively. The coupling efficiency between the focused laser Gaussian beam and the waveguide mode can be computed using equation (3.9a), where n_0 and n_{WG} are the refractive index values of free space and the waveguide mode, respectively. The coupling efficiency depends on the phase-matching efficiency and the degree of field overlap, which can be separated into equations (3.9b) and (3.9c). Conceptually, the phase-matching efficiency is similar to the transmittance at an interface, which can be increased to 100% by coating an AR layer onto the end surface of a dielectric

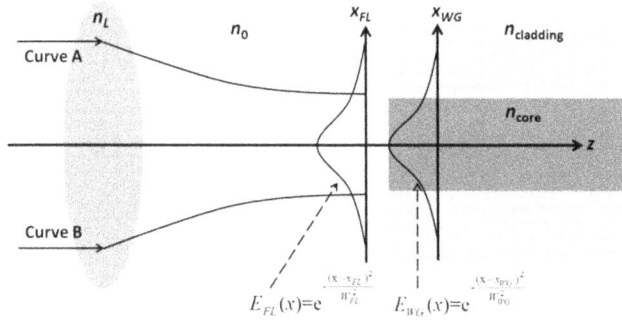

Figure 3.8. One-dimensional modal profiles of a focused laser Gaussian beam and a dielectric waveguide.

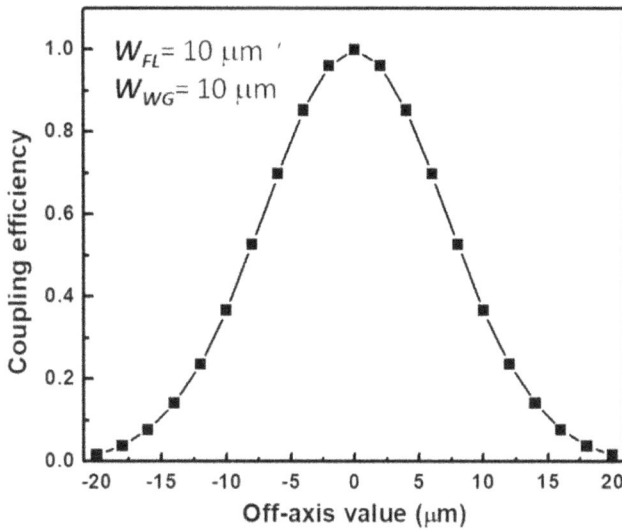

Figure 3.9. Coupling efficiency between a focused laser Gaussian beam and the waveguide mode for different off-axis values.

waveguide. The degree of field overlap can reach 100% when x_{FL} and W_{FL} are equal to x_{WG} and W_{WG}, respectively. However, imperfect optical alignment can reduce the coupling efficiency between the focused laser Gaussian beam and the dielectric waveguide mode. To illustrate the effect of off-axis coupling, the x-position-dependent coupling efficiency is plotted in figure 3.9. Insertion loss (IL) is widely used as a metric of coupling loss, which is given by equation (3.9d); the unit of IL is dB. For example, the calculated IL value is 3.01 dB when the coupling efficiency is 0.5. Therefore, 3 dB is widely used as a reference value when evaluating the performance of optical components, which can be used to determine the acceptable tolerance in optical alignment in the x-direction (ΔX_{3dB}). When W_{FL} and W_{WG} are both 10 μm, the ΔX_{3dB} value is about ±8 μm.

$$\eta = \frac{\mathrm{Re}(n_{WG})}{n_0} \left| \frac{2n_0}{n_0 + n_{WG}} \right|^2 \frac{\left[\int_{-\infty}^{\infty} E_{FL}(x) E_{WG}^*(x)\mathrm{d}x \right]^2}{\int_{-\infty}^{\infty} E_{FL}(x) E_{FL}^*(x)\mathrm{d}x \int_{-\infty}^{\infty} E_{WG}(x) E_{WG}^*(x)\mathrm{d}x} \qquad (3.9a)$$

$$\eta_{\mathrm{phase}} = \frac{\mathrm{Re}(n_{WG})}{n_0} \left| \frac{2n_0}{n_0 + n_{WG}} \right|^2 \qquad (3.9b)$$

$$\eta_{\mathrm{field}} = \frac{\left[\int_{-\infty}^{\infty} E_{FL}(x) E_{WG}^*(x)\mathrm{d}x \right]^2}{\int_{-\infty}^{\infty} E_{FL}(x) E_{FL}^*(x)\mathrm{d}x \int_{-\infty}^{\infty} E_{WG}(x) E_{WG}^*(x)\mathrm{d}x} \qquad (3.9c)$$

$$IL = -10 \log_{10} \eta \qquad (3.9d)$$

3.5 Plasmonic lenses

Surface plasmon polariton (SPP) waves can be supported at an air/metal interface, as shown in figure 3.10(a). The electric fields in the air and metal regions are described by equations (3.10a) and (3.10b), respectively, which show that SPP waves propagate along the air/metal interface. Here, α_a and α_m are the decay coefficients in the air and metal regions, respectively, which indicate that the SPP wave is a bound optical mode that is confined to the x-direction. The electric displacements are continuous at the interface ($x = 0$), which can be written as equation (3.10c). According to the prediction of the lossless Drude model, the relative permittivity of a metal is negative when the operating frequency is lower than the plasma frequency. Therefore, the value of E_{air} is negative (positive) when the value of E_{metal} is positive (negative). The directions of the electric fields indicate that electrons and holes are locally modulated at the metal surface by electric fields, as shown in figure 3.10(b). In other words, SPP waves originate from the mutual interactions between time-varying electric fields and electrons at the air/metal interface, which is a macroscopic result. The modal field distribution of the SPP wave in the x-direction implies that the photons are mainly concentrated at the air/metal interface, thereby resulting in a slower propagation speed. At a dielectric/metal interface, the two decay constants

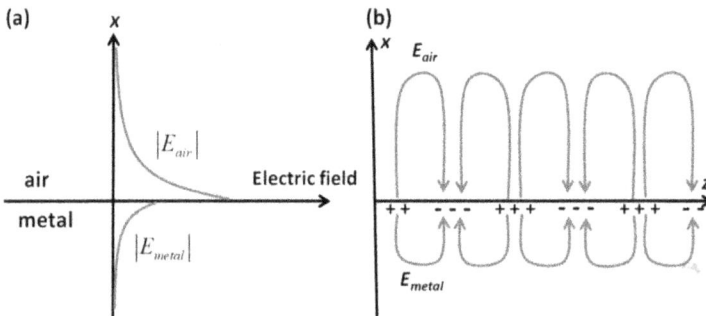

Figure 3.10. (a) Electric field distribution of an SPP wave supported by an air/metal interface. (b) Relation between the electric fields and charges at an air/metal interface.

(α_d and α_m) are related to the material properties, which can be written as equations (3.10d) and (3.10e), where ε_d, ε_m, and k_{spp} are the relative permittivity of the dielectric, the relative permittivity of the metal, and the propagation constant of the SPP wave, respectively. In general, $1/\varepsilon_d$ ($1/\varepsilon_m$) is far smaller than $\sqrt{k_{spp}^2 - \varepsilon_d(\omega^2/c^2)}$. $\left(\sqrt{k_{spp}^2 - \varepsilon_m(\omega^2/c^2)}\right)$. Therefore, α_d and α_m can be approximated as $\sqrt{k_{spp}^2 - \varepsilon_d(\omega^2/c^2)}$ and $\sqrt{k_{spp}^2 - \varepsilon_m(\omega^2/c^2)}$, respectively. To solve the dispersion relation of the SPP wave, conditions of continuous E_z and H_y at the dielectric/metal interface must be used, which can be written as equations (3.10f). By combining equations (3.10d), (3.10e), and (3.10f), the k_{SPP} of a dielectric/metal interface can be computed using equation (3.10g) if ε_d, ε_m, and ω are known. The modal index of the SPP wave is denoted by n_{SPP} in equation (3.10g).

$$\vec{E}_{air} = E_{air}e^{i(k_{spp}z-\omega t)} \times e^{-\alpha_a x}\hat{x} \tag{3.10a}$$

$$\vec{E}_{metal} = E_{metal}e^{i(k_{spp}z-\omega t)} \times e^{\alpha_m x}\hat{x} \tag{3.10b}$$

$$D_{air} = \varepsilon_0\varepsilon_{air}E_{air} = D_{metal} = \varepsilon_0\varepsilon_{metal}E_{metal} \tag{3.10c}$$

$$\alpha_d = \frac{-(1/\varepsilon_d) + \sqrt{4(k_{spp}^2 - \varepsilon_d(\omega^2/c^2)) + (1/\varepsilon_d^2)}}{2} \sim \sqrt{k_{spp}^2 - \varepsilon_d(\omega^2/c^2)} \tag{3.10d}$$

$$\alpha_m = \frac{(1/\varepsilon_m) + \sqrt{4(k_{spp}^2 - \varepsilon_m(\omega^2/c^2)) + (1/\varepsilon_m^2)}}{2} \sim \sqrt{k_{spp}^2 - \varepsilon_m(\omega^2/c^2)} \tag{3.10e}$$

$$\frac{\alpha_d}{\alpha_m} = \frac{\varepsilon_d}{\varepsilon_m} \tag{3.10f}$$

$$k_{spp} = \frac{\omega}{c}\sqrt{\frac{\varepsilon_d\varepsilon_m}{\varepsilon_d + \varepsilon_m}} = \frac{\omega}{c}n_{spp} \tag{3.10g}$$

Plasmonic components have been investigated in the optical communication band centered at $\lambda_0 = 1550$ nm in order to realize OICs. At $\lambda_0 = 1550$ nm, the real part and imaginary part of the relative permittivity (ε_R and ε_I) of Au are -115 and 11.3, respectively. To simplify the calculations, the ε_I of Au is ignored in the following discussion. Therefore, the calculated n_{SPP}, α_d, and α_m values of an air/Au interface at $\lambda_0 = 1550$ nm are 1.004, 3.629×10^5 m^{-1}, and 4.366×10^7 m^{-1}, respectively. When the electric field decays to 1/e in the x-direction, the corresponding length is defined as the penetration depth (L_D), i.e. $e^{-\alpha L_D} = e^{-1}$. At $\lambda_0 = 1550$ nm, the L_D values of the SPP wave in the air and Au regions are 2756 and 23 nm, respectively, which indicates that the SPP wave is localized at the wavelength scale. When ε_d is increased from one to four, the L_D value of the SPP wave in the dielectric region is decreased from 2756 to 1695 nm, which indicates that the field localization of the SPP waves can be increased by increasing the relative permittivity of the dielectric medium.

The ohmic loss of SPP waves limits the application of plasmonic components in OICs. Therefore, it is important to know the propagation distance of the SPP waves. When the ε_I of Au is considered in equation (3.10g), the Re$[n_{spp}]$ and Im$[n_{spp}]$ values of the air/Au interface are 1.009 and 8.610×10^{-4}, respectively. At an air/Au interface, the electric field of an infinite SPP wave can be written as equation (3.11a), where the imaginary part of the propagation constant (Im$[k_{spp}]$) is defined as the attenuation coefficient (α_z), which is 3.490×10^3/m. When the intensity of the SPP wave decays to 0.5 in the z-direction, the corresponding length is defined as the propagation length (L_P), i.e. $e^{-2\alpha_z L_P} = 0.5$. The relation between L_P and α_z can be written as equation (3.11b). When α_z is 3.490×10^3/m, L_P is 99.3 μm. The ratio of L_P to λ_{spp} can be used as a metric in the evaluation of plasmonic waveguide-based devices, such as ring resonators, Bragg reflectors, and coupled arrays.

$$\vec{E}_{spp} = E_{spp}(x)e^{-i\omega t}e^{i\,\mathrm{Re}[k_{spp}]z}e^{-\mathrm{Im}[k_{spp}]z} \tag{3.11a}$$

$$L_P = \frac{\ln(2)}{2}\alpha_z^{-1} \tag{3.11b}$$

Equations (3.10d), (3.10e), and (3.10g) show that the modal size (penetration depth) is inversely proportional to the modal index of the SPP waves, which means that the modal index of the SPP waves can be manipulated by varying the field distribution through the use of different plasmonic structures. The modal index of a metal/dielectric/metal (MDM) plasmonic waveguide is inversely proportional to the thickness of the dielectric layer, which can be used to manipulate the phase at the end of the plasmonic waveguide and thereby form a lens device, as shown in figure 3.11. Here, L, W_G, and T_m are the length, width, and metal thickness of the

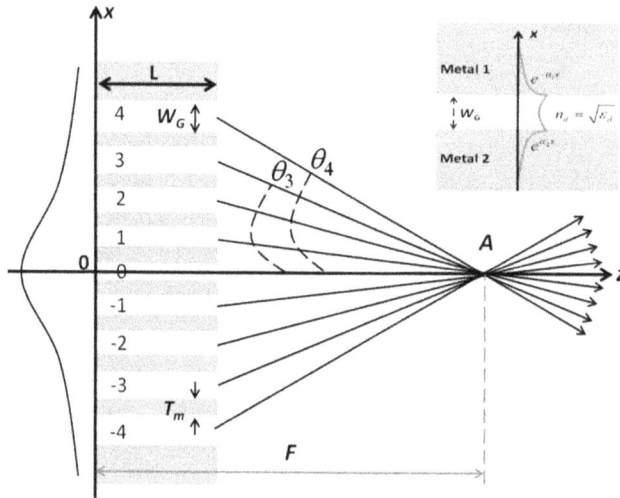

Figure 3.11. Optical paths in a metal/dielectric/metal waveguide array-based lens. The inset shows the field distribution.

MDM waveguide, respectively. The T_m value must be larger than the penetration depth of the metal to avoid unwanted interference between adjacent MDM waveguides. After exciting SPP waves in nine MDM plasmonic waveguides, a collimated and E_x-directed Gaussian laser beam can be concentrated at point A by adjusting the phase distribution at the end surface. A narrower width (W_G) corresponds to a larger modal index, which results in a longer optical path length in the MDM plasmonic waveguide. The optical path length (OPL) of the waveguide equals $n_{spp}L$. In waveguide i, the total OPL from the left end surface of the array to point A can be written as equation (3.12a), where i takes values of $-4, -3, -2, -1, 0, 1, 2, 3$, or 4. The OPL must be a fixed value for the nine optical paths, thereby forming constructive interference at point A, which is the focal point of the plasmonic lens. The difference in the modal index values of adjacent MDM waveguides is related to the focusing angle, which can be written as equation (3.12b), where i takes values of 0, 1, 2, or 3.

$$\text{OPL} = n_{\text{spp},\, i}L + n_{\text{air}}(F - L)/\cos \theta_i \tag{3.12a}$$

$$n_{\text{spp},\, i+1} - n_{\text{spp},\, i} = (F - L)\left(\frac{\cos \theta_{i+1} - \cos \theta_i}{\cos \theta_{i+1} \cos \theta_i} \right) \tag{3.12b}$$

The modal index of an MDM plasmonic waveguide can be computed by solving the transcendental equation given in [5] using a graph method. The transcendental equation for κ is written as equation (3.13a), where κ is the propagation constant in the x-direction, while α_1 and α_2 are the decay constants in metals 1 and 2 of the MDM plasmonic waveguide, respectively. The two decay constants are related to the refractive indices of the two metals, which are given by equations (3.13b) and (3.13c), where λ_0 is the wavelength in free space, while p and q are $\varepsilon_d/\varepsilon_{m1}$ and $\varepsilon_d/\varepsilon_{m2}$, respectively, where the subscript $m1$ ($m2$) represents metal 1 (metal 2). For simplicity, $(p\alpha_1 + q\alpha_2)/2$ is defined as S. At $\lambda_0 = 1550$ nm, the real and imaginary parts of the relative permittivity (ε_R and ε_I) of Au are -115 and 11.3, respectively. To simplify the calculations, the ε_I of Au is ignored in the calculation of the transcendental equation. Figure 3.12 plots the curves Y_1 and Y_2 of the transcendental equation of an MDM plasmonic waveguide for different W_G values. The intersection of curves Y_1 and Y_2 gives the solution of κ (this approach is known as a graph method). The κ values are 2.765×10^6 m^{-1} and 3.910×10^6 m^{-1} for WG values of 100 and 50 nm, respectively. The modal index of an MDM plasmonic waveguide can be computed using equation (3.13d). The calculated n_{spp} value significantly increases from 1.297 to 1.402 as the W_G value decreases from 100 to 50 nm. In other words, equations (3.12a) and (3.13a) can be used to design plasmonic lenses at subwavelength scales.

$$\kappa^2 + pq\alpha_1\alpha_2 = -2S\kappa \coth(\kappa W_G) \tag{3.13a}$$

$$\alpha_1 = \sqrt{k_0^2(\varepsilon_d - \varepsilon_{m1}) + \kappa^2} \tag{3.13b}$$

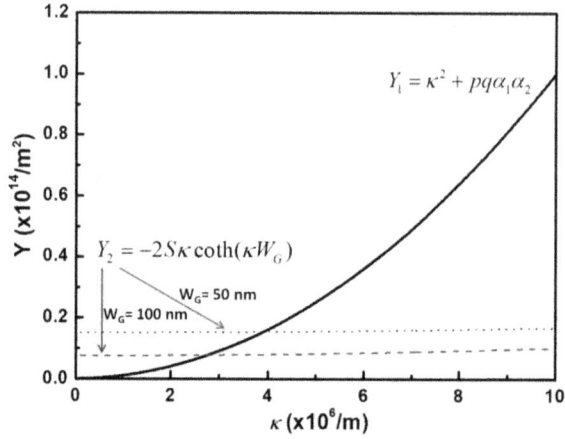

Figure 3.12. Curve of the transcendental equation for an MDM plasmonic waveguide. The dielectric layer and the metal used in the calculation were free space ($\varepsilon_d = 1$) and Au, respectively.

$$\alpha_2 = \sqrt{k_0^2(\varepsilon_d - \varepsilon_{m2}) + \kappa^2} \qquad (3.13c)$$

$$n_{spp} = \sqrt{\varepsilon_d + (\kappa/k_0)^2} \qquad (3.13d)$$

Bibliography

[1] Tan C Z 1998 Dtermination of refractive index of silica glass for infrared wavelengths by IR spectroscopy. *J. Non-Cryst. Solids* **223** 158–63
[2] Schott Zemax Catalog 2017-01-20b
[3] Li H H 1980 Refractive index of alkaline earth halides and its wavelength and temperature derivatives *J. Phys. Chem. Ref. Data* **9** 161–289
[4] Polyanskiy M N 2024 Refractiveindex.info database of optical constants *Sci. Data* **11** 94
[5] Kekatpure R D, Hryciw A C, Barnard E S and Brongersma M L 2009 Solving dielectric and plasmonic waveguide dispersion relations on a pocket calculator *Opt. Express* **17** 24112–29

IOP Publishing

Light–Material Interactions and Applications in Optoelectronic Devices

Anjali Chandel and Sheng Hsiung Chang

Chapter 4

Optical spectroscopic techniques

The first part of this chapter describes the multifunctional optical microscope (OM)-based spectrometer. The functions of the mirror, half-wave plate (HWP), polarization beam splitter (PBS), objective lens, flip mirror (FM), linear polarizer, filter, coupler/fiber/connector, and grating components are discussed. The importance and reason for the alignment of the laser beam with the optical axis of the objective lens are graphically explained. In the second part, the working mechanisms and applications of transmittance/reflectance, photoluminescence, Raman scattering, and scanning near-field optical microscopy (SNOM) spectrometers are described. In scattering-type SNOM spectrometry, the dipole model and effective polarizability model are used to compute the scattering cross-section coefficient, which can be used to determine the dielectric permittivity values of nanocomposite materials at the nanoscale and thereby map the two-dimensional spatial distributions of optical responses.

4.1 Optical-microscope-based spectroscopy

Chapters 1 and 3 explained the properties and working mechanisms of linear polarizers, wave plates (WPs), dielectric lenses, beam expanders, objective lenses, optical couplers, and plasmonic lenses; this represents the background knowledge required to construct home-made OM-based spectroscopes. Figure 4.1 schematically illustrates the optical layout of a multifunctional OM-based spectroscope.

The light source can be a laser or a lamp. The beam produced by a lamp must be collimated in order to be guided and controlled by the subsequent optical components. However, the light waves emitted by a lamp are incoherent, which results in poor collimation in the propagation path. Therefore, the optical path length from the light source to objective 1 (OB1) is suggested to be shorter than 2 m.

doi:10.1088/978-0-7503-6099-9ch4

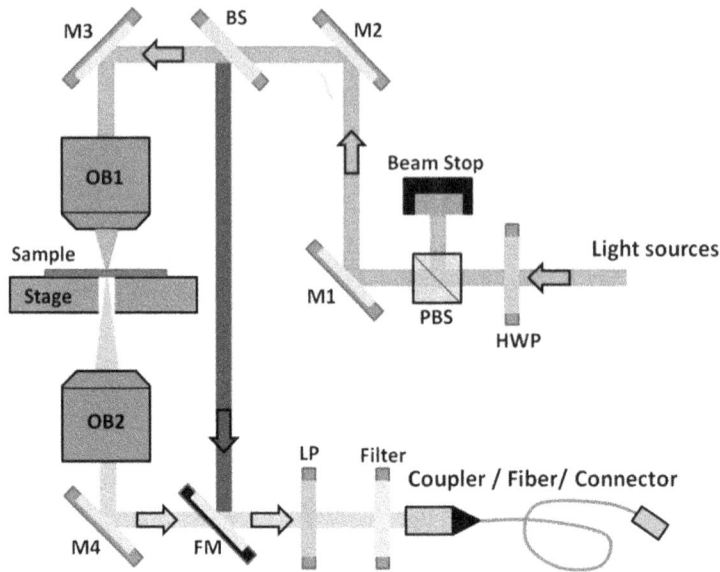

Figure 4.1. Optical layout of a multifunctional OM-based spectroscope. HWP: half-wave plate; PBS: polarization beam splitter; M: mirror; BS: beam splitter; OB: objective; FM: flip mirror; LP: linear polarizer.

Fortunately, laser-induced white emission sources have high coherence [1] and have been used in many commercial optical spectrometers.

If the light source is a linearly polarized laser, the combination of a PBS and an HWP can be used to vary the power of the transmitted laser beam. By rotating the ϕ angle of the WP, the polarization state of the laser beam can be changed from s polarization to p polarization. Upon passing through the PBS, the s-polarized and p-polarized laser beams are mainly reflected and transmitted, respectively. To eliminate the influence of unwanted light on OM-based spectroscopes, a beam stop must be used. The inner structure of a beam stop is similar to a black cone, which produces multiple scatterings and reabsorptions. Mirrors 1, 2, 3, and 4 are used to reflect the light beams.

When the operating wavelength ranges from 400 to 1000 nm, protected Ag films coated onto flat substrates are widely used as mirrors. In general, a thin and dense dielectric film is coated onto the Ag/glass substrate in order to minimize the formation of silver oxides at the surface of the Ag film. However, the reflectance of protected Ag mirrors is lower than that of Ag mirrors because of the longer penetration depth in the Ag film. It should be noted that the reflectance of a protected Ag mirror is higher when the incident light beam is in an s-polarization state. In other words, the penetration depth of a p-polarized light beam is longer than that of an s-polarized beam in an Ag film, which is related to ohmic loss. At $\lambda_0 = 1550$ nm, the refractive index values of Ag, Au, Al, and Cu are $0.144 + i11.366$, $0.524 + i10.742$, $1.579 + i15.658$, and $0.716 + i10.655$, respectively. The reflectance of a metal film can be computed using the simple relation: $R = |(n_{air} - n_{metal})/(n_{air} + n_{metal})|^2$, where n_{air} and n_{metal} are the refractive index values of air and metal, respectively. The reflectance values of air/Ag, air/Au, air/Al, and air/Cu interfaces are 99.56%, 98.22%, 97.49%, and 97.54%, respectively, at normal incidence, which explains why Ag films are widely used as mirrors.

Beam splitters (BSs) are widely used in reflection-type OM-based spectroscopes. In general-purpose spectroscopy, the transmittance/reflectance (T/R) ratio of the BS is one in order to excite the sample and collect the reflected optical signal effectively and simultaneously. To confirm that the optimal T/R ratio of the BS is one, the intensity of the measured signal is computed using the simple relation: $I_{measured} = T_{BS} \times R_{sample} \times (1 - T_{BS})$, where T_{BS} is the transmittance of the BS, R_{sample} is the reflectance, and $(1 - T_{BS})$ is the reflectance of the BS. The relation is a parabolic function of T_{BS}, which reaches its maximum value when the derivative of $I_{measured}$ equals zero, i.e. $dI_{measured}/dT_{BS} = 0$. The calculated maximum $I_{measured}$ is $0.25 R_{sample}$ when T_{BS} is 0.5, which confirms that the optimal T/R ratio of the BS is one.

In reflection-type OM-based spectroscopes, the excitation laser beam must be aligned with the optical axis of OB1 in order to efficiently collect backscattered light such as photoluminescence (PL) and Raman scattering from the illuminated region of the sample. In general, the diameter of the collimated laser beam is smaller than the entrance aperture stop of OB1. Therefore, the focusing angle of the focused laser beam is smaller than the collection angle of OB1. Figure 4.2(a) plots the optical paths of a laser beam and the collected light when the laser beam is aligned with the optical axis (vertical dashed line) of the lens, where curves A and B are the waists of the laser beam, and the blue dashed lines are determined by the edges L and R of the entrance aperture stop of the lens. When the laser beam is not aligned with the optical axis of the lens, the emitted light is partially blocked by the fixed entrance/exit aperture stop, which shows that the optical alignment of the laser beam with a lens or an objective is critically important in order to effectively collect the backscattered PL and Raman scattering from the illuminated region of the sample.

In transmission-type OM-based spectroscopes, forward-scattered PL and Raman scattering can be collected by OB2. When the collected light comes from M4 or the BS, the FM must be flipped down or up, respectively. When the FM is flipped down (up), forward-scattered (backscattered) light can be collected by the optical coupler.

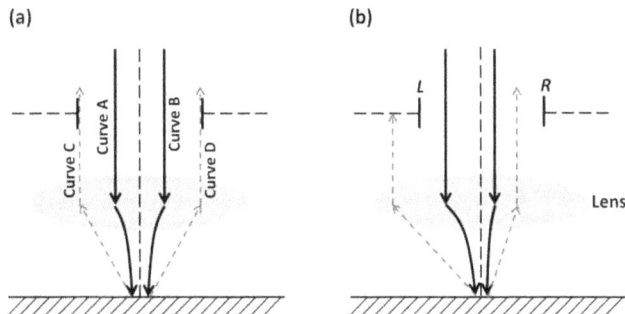

Figure 4.2. Optical paths of the laser light and the collected light. (a) Laser beam aligned with the optical axis. (b) Laser beam unaligned with the optical axis.

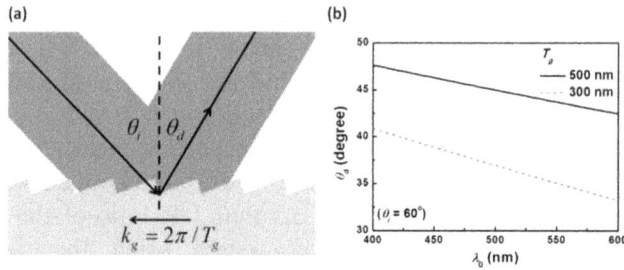

Figure 4.3. (a) Optical diffraction produced by an asymmetric grating; (b) two $\lambda_0 - \theta_d$ curves.

The linear polarizer (LP) is used to analyze the polarization state of the light scattered by the illuminated region of the sample. The filter is used to block the residual excitation laser light. The filter can be a long-pass filter or a band-stop filter. The optical coupler is connected to the left-hand end of the optical fiber. The length of the optical fiber can be longer than 1 m, which reduces background noise from the environmental lighting. The connector at the right-hand end of the optical fiber is used to connect an optical spectrometer which consists of a slit, gratings, and a charge-coupled device (CCD) array. To achieve optimal wavelength resolution, the width of the slit and the pixel size of the CCD array should be similar. In general, the pixel width of a one-dimensional CCD linear array ranges from 40 to 10 μm, which determines the narrowest width of the slit. However, the core diameters of the widely used multimode optical fibers range from 100 to 300 μm in order to effectively collect the scattered light from the sample. In other words, the optimal slit width is typically about 100 μm due to the trade-off between light intensity and wavelength resolution.

Gratings are used to spatially separate light of different wavelengths, as shown in figure 4.3(a), where θ_i and θ_d are the incident angle of the collimated light beam and the related diffraction angle, respectively. The diffraction angle can be computed using the simple relation: $k_0 \sin \theta_i = k_0 \sin \theta_d + k_g$, where k_0 and k_g are the propagation constant of the light beam and the momentum of the grating, respectively. When a collimated broadband light beam is incident on the asymmetric grating, θ_i and k_g ($2\pi/T_g$) have fixed values, which results in a wavelength-dependent diffraction angle: $\theta_d(\lambda_0) = \sin^{-1}[\sin \theta_i - k_g \lambda_0/(2\pi)]$. Figure 4.3(b) plots the $\lambda_0 - \theta_d$ curves for θ_i fixed at 60°, which shows that a shorter grating period (T_g) results in a larger angular dispersion. The angular dispersion can be computed using the simple relation: $d\theta_d/d\lambda_0 = -1/(T_g \cos \theta_d)$. On the other hand, the T_g value can be used to compute the number of grooves in 1 mm using the simple relation: $D_g = 1 \text{ mm}/T_g$. When the T_g value is 500 nm (300 nm), the D_g value is 2000 lines (3333 lines). In other words, the use of a higher-density grating results in better wavelength resolution of the optical spectrum analyzer.

4.2 Transmittance/reflectance spectrometry

When a low-coherence light source is used in a transmittance/reflectance spectrometer, the thin-film samples should be located in the focal plane of a lens in order to fix

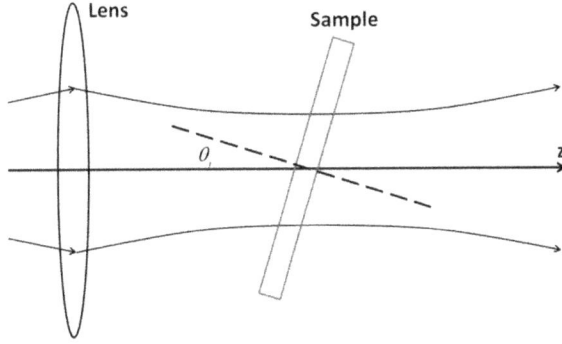

Figure 4.4. The sample is illuminated with a tilt angle.

the incident angle in the illumination region, as shown in figure 4.4. When the Rayleigh range is far larger than the thickness of the thin-film sample, the incident angle is denoted by θ_i. In other words, the incident angle can be varied by tilting the sample. When the focal length of the lens is far larger than the thickness of the thin-film sample, the Rayleigh range is insensitive to the wavelength. In general, the focal length of the lens is larger than 10 cm. By combining equations (3.2f) and (3.4a), the Rayleigh range of a lens can be computed using equation (4.1), where λ_0 is the wavelength of free space, w is the beam radius at the lens, and F is the focal length of the lens. When λ_0, w, and F are 1550 nm, 5 mm, and 10 cm, the calculated value of $2z_R$ equals 394 μm, which is far larger than the thickness of the film. In other words, multiple interference in thin films is not influenced by wave diffraction effects. The transmitted light waves do not have to be completely collected to determine the transmittance at normal incidence. The transmittance of the sample can be computed using the simple relation: $T = P_{\text{sample}}/P_{\text{air}}$. Here, P_{sample} is the measured power when the sample is placed in the focal plane, while P_{air} is the measured power when the sample is removed. In a uniform thin-film sample, transmittance values are also uniform.

$$z_R = (\pi/\lambda_0)\left(\frac{w^2 - \sqrt{w^4 - 4F^2\lambda_0^2/\pi^2}}{2} \right) \tag{4.1}$$

Figure 4.5 plots the transmittance spectrum of an indium tin oxide (ITO)/glass substrate and the absorbance spectrum of a $CH(NH_2)_2PbI_3$ (FAPbI$_3$)/glass substrate. ITO and FAPbI$_3$ are a transparent conductive material and a light-absorbing material, respectively. In the transmittance spectrum of an ITO/glass sample, the phenomena of thin-film interference and the free carrier effect are observed. The thin-film interference results in ripples in the transparent wavelength range from 450 to 800 nm.

The three labeled wavelength values λ_A, λ_B, and λ_C shown in figure 4.5(a) can be used to determine the refractive index (thickness) by applying equations (4.2a) and (4.2b) when the thickness (refractive index) of the ITO film is known. L is the

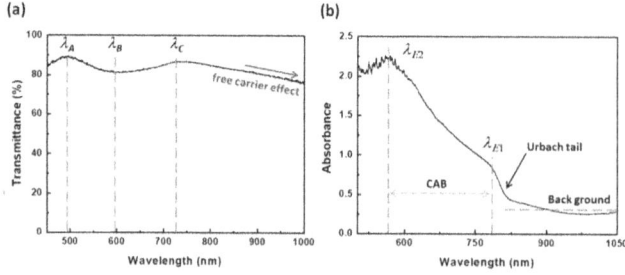

Figure 4.5. (a) Transmittance spectrum of an indium tin oxide (ITO)/glass substrate. (b) Absorbance spectrum of a $CH(NH_2)_2PbI_3$ (FAPbI$_3$)/glass substrate.

thickness of the ITO film. The value of λ_P can equal λ_A or λ_C. It should be noted that the refractive index (n) is assumed to be a fixed value in the wavelength range from λ_A to λ_C. Equation (4.2a) can be proved using the two antireflective coating conditions: $4n_C L = (1 + 2l)\lambda_C$ and $4n_A L = (3 + 2l)\lambda_A$, where n_C (n_A) is the refractive index of the thin film at λ_C (λ_A) and l is an integer. When l is zero, the L value of the ITO thin film equals a quarter wavelength, which means that the thin film can be used as an antireflective coating layer. The difference between λ_C and λ_A equals $4L[(3n_C - n_A) + 2l(n_C - n_A)]/[(1 + 2l)(3 + 2l)]$. When n_C equals n_A, λ_A/λ_C equals $(1 + 2l)/(3 + 2l)$. The difference between λ_C and λ_A can then be rewritten as $(\lambda_C\lambda_A)/(2Ln_C)$, which satisfies equation (4.2a).

$$nL = \frac{\lambda_C \lambda_A}{2(\lambda_C - \lambda_A)} \tag{4.2a}$$

$$nL = \frac{\lambda_B \lambda_P}{4 \, |\lambda_B - \lambda_P|} \tag{4.2b}$$

The λ_A, λ_B, and λ_C values are about 493, 596, and 735 nm, respectively. The refractive index of ITO is 1.864 at $\lambda_0 = 550$ nm [2]. The L value of the ITO thin film is computed to be about 382.6 nm when λ_A and λ_B are used in equation (4.2b). The calculated refractive index is about 2.059 when λ_B and λ_C are used in equation (4.2b); this value is larger than the experimentally reported value of 1.765 at $\lambda_0 = 650$ nm [2]. This means that the free carrier effect results in a shorter wavelength at λ_C. The free carrier effect can be understood by analyzing the Drude model shown in equation (2.8h). After a simple rearrangement, the frequency-dependent relative permittivity can be rewritten as equation (4.3a). When the angular frequency ($\omega = 2\pi f$) is higher than the plasma frequency (ω_p) and the collision frequency ($1/\tau$), the imaginary part of the relative permittivity can be written as equation (4.3b), which shows that the attenuation of light waves in a conductive material is inversely proportional to the frequency (f), thereby resulting in the free carrier effect in the transmittance spectrum.

$$\varepsilon_r(\omega) = \left(1 - \frac{\omega^2 \tau^2 \omega_p^2}{\tau^2 \omega^4 + \omega^2}\right) + i\frac{\omega \tau \omega_p^2}{\tau^2 \omega^4 + \omega^2} \tag{4.3a}$$

$$\text{Im}[\varepsilon_r] \sim \frac{\tau \omega_p^2}{\omega} \tag{4.3b}$$

In figure 4.5(b), the absorbance (A) is computed using the simple relation: $A = -\log_{10}(T)$, where T is transmittance. The T value ranges from 0% to 100%, resulting in an A value that ranges from infinity to zero. In other words, the absorbance must be positive. The features of the absorbance curve are mainly related to light absorption and thin-film interference. In the wavelength range from 450 to 800 nm, the two peaks (λ_{E1} and λ_{E2}) and one stepped slope can be assigned to exciton absorptions and the continuous absorption band (CAB), respectively. At the band edge, the absorbance curve is defined as the Urbach tail, which is related to the crystallinity of light-absorbing materials. A sharper Urbach tail corresponds to better crystallinity. The Urbach energy can be obtained by fitting the Urbach tail to an exponential decay function which is written as equation (4.4), where A_0 is related to the absorbance strength, E_g is the absorption bandgap, E_U is the Urbach energy, and A_b is the background value of the absorbance spectrum [3]. Figure 4.6(a) plots the absorbance spectrum of a FAPbI$_3$/glass sample in the energy range from 1.18 to 1.64 eV. The average values of A_b are A_1 and A_2, which are related to the destructive and constructive thin-film interferences of the transmission, respectively. It should be noted that the fitting range of photon energy must be located in the range of exponential decay (E_1 and E_2) in order to obtain the correct Urbach energy of a light-absorbing material. Figure 4.6(b) shows that the absorbance curve is a good fit for an exponential decay function in the photon energy range from 1.52 to 1.55 eV when A_b is fixed at 0.3. The calculated E_g and E_u values are 1.524 eV and 41.1 meV, respectively, which shows that a thin film of FAPbI$_3$ can be used as an efficient light-absorbing material in light-emitting diode and photodetector applications.

$$A(E) = A_b + A_0 e^{-\left(\frac{E - E_g}{E_U}\right)} \tag{4.4}$$

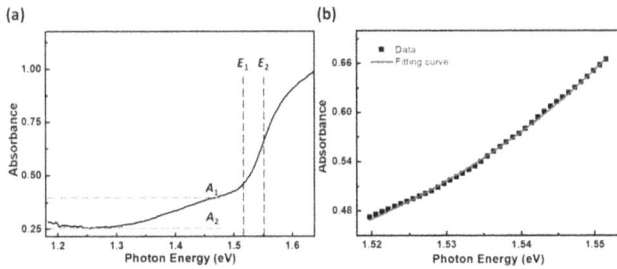

Figure 4.6. (a) Absorbance spectrum of an FAPbI$_3$/glass sample. (b) Absorbance curve in the exponential decay range and the fitted curve.

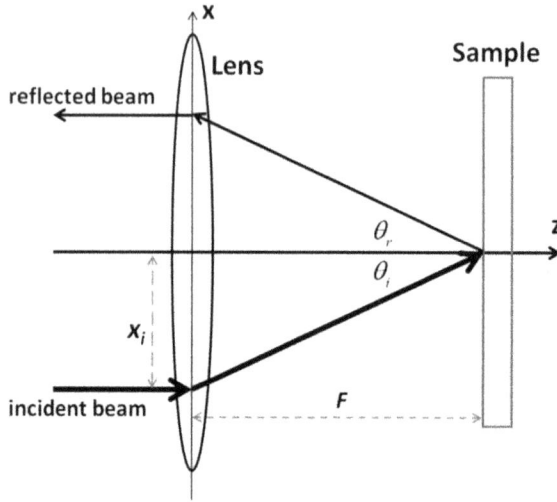

Figure 4.7. An optical configuration for measuring the incident angle-dependent reflectance.

Figure 4.7 plots an optical configuration that can be used to measure the incident angle-dependent reflectance when the focal length is far larger than the diameter of the incident laser beam. The incident angles θ_i are related to x_i and F according to the geometric relation $\tan \theta_i = x_i/F$. When the focal length (F) is fixed, the x_i value determines the θ_i value. According to Snell's law of reflection, θ_r equals θ_i. Therefore, the reflected beam is antiparallel to the incident beam when the normal surface is aligned with the optical axis of the lens. When F and the clear aperture of the lens are 5 and 2 cm, respectively, the incident angle ranges from zero to $\tan^{-1}(1/5)$. The diameter of the incident Gaussian laser beam should be far smaller than the clear aperture of the lens in order to obtain higher angular resolution. When the diameter of the incident beam and F are 1 mm and 5 cm, respectively, the angular resolution equals $\tan^{-1}(1/50)$, which is about $1.15°$.

At the band edge, the excitonic feature of a semiconductor thin film can be observed in the reflectance spectrum without a significant influence from the thin-film interference effect. Semiconductors have one excitonic transition at the band edge, which can be illustrated using one Lorentz oscillator. The relative permittivity of one Lorentz oscillator is written in equation (2.5g). After a simple rearrangement, the frequency-dependent relative permittivity can be rewritten as equation (4.5), where ε_∞ is the relative permittivity at an infinite frequency, w_c is the characteristic angular frequency, w_0 is the oscillation frequency, and Γ is the collision frequency. w_c is related to the density of bound electrons. Figure 4.8 plots the real and imaginary parts of one Lorentz oscillator when ε_∞, w_c, w_0, and Γ are 4, 5×10^{14}/s, $2\pi f_o$, and $2\pi f_o/100$, respectively. The oscillation frequency is computed using the simple relation $f_o = c/\lambda_0$, where λ_0 equals 700 nm. When the wavelength of the incident photons is longer than the peak wavelength of the dashed curve, there is a sharp peak in the real part of the permittivity. In the sharp peak wavelength range, the incident photons can be absorbed, thereby reducing the thin-film interference

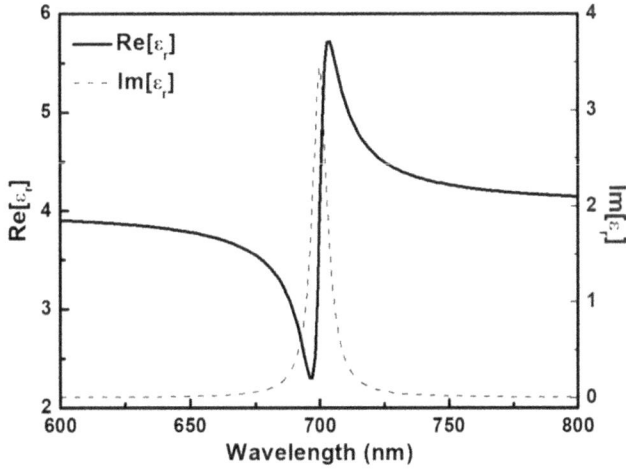

Figure 4.8. Real and imaginary parts of the relative permittivity of one Lorentz oscillator.

feature. In other words, it is predicted that there is a sharp peak in the reflection spectrum, which is mainly related to the excitonic feature of the semiconductor thin film.

$$\varepsilon_r(\omega) = \left[\varepsilon_\infty + \frac{\omega_c^2(\omega_o^2 - \omega^2)}{(\omega_o^2 - \omega^2)^2 + \Gamma^2\omega^2}\right] + i\left[\frac{\omega_c^2\Gamma\omega}{(\omega_o^2 - \omega^2)^2 + \Gamma^2\omega^2}\right] \qquad (4.5)$$

4.3 Photoluminescence/Raman scattering spectrometry

When the photon energy of light waves is larger than the bandgap of a semiconductor, the relaxation processes of photoexcited electrons can emit Raman scattering and PL, which are related to molecular and crystal structures. Figure 4.7 shows an optical configuration that is widely used to measure the backward PL and Raman scattering signals when the incident angle (θ_i) is zero. In a photoexcited semiconductor, electrons can be excited from the ground state to the excited state when the photon energy of the light wave is larger than the absorption bandgap, which forms hot electrons (see figure 4.9). In organic (inorganic) materials, the lifetime of the hot electrons is about 100 fs (1 ps). Coherent collisions between hot electrons and phonons (molecular vibrations) result in weak Raman scattering intensities owing to the large frequency mismatch. After the thermalization process is complete, the hot electrons relax to the metastable state of the conduction band. In the thermalization process, the excited electrons transfer a portion of their energy to the lattice via incoherent collisions, thereby resulting in an increase in the temperature of the excited semiconductor. In the metastable state, the electrons are called cold electrons, which have a lifetime of about 100 ps (10 ns) in organic (inorganic) materials. When cold electrons radiatively relax to the ground state, the photon energy of the PL is lower than the absorption bandgap because of the Stokes shift. In general, a larger Stokes shift corresponds to larger Urbach energy, which

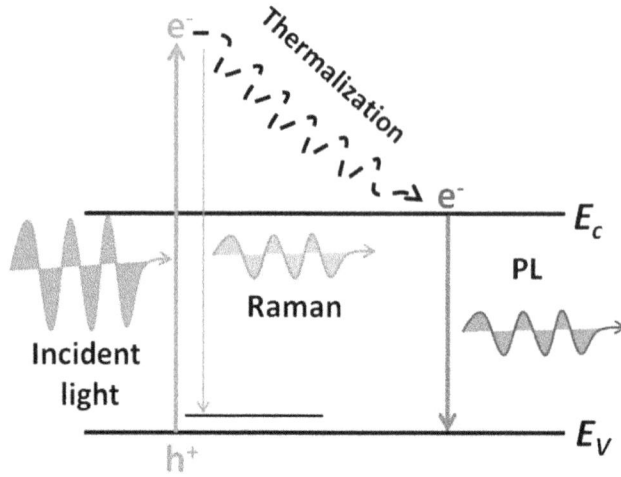

Figure 4.9. Radiative relaxation processes of photoexcited electrons in a semiconductor.

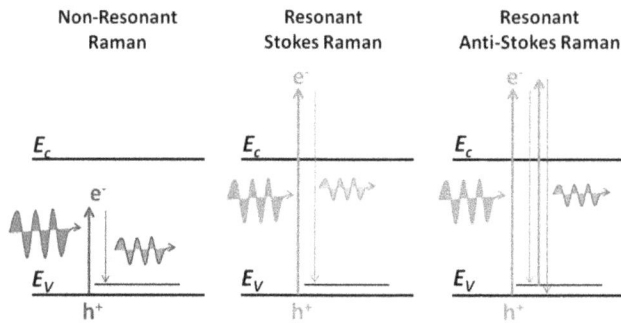

Figure 4.10. Energy diagrams of non-resonant Raman scattering, resonant Stokes Raman scattering, and resonant anti-Stokes Raman scattering.

follows the relation $\Delta_{\text{Stokes}} = E_U/2$, where Δ_{Stokes} is the Stokes shift and E_U is the Urbach energy. In other words, the Stokes shift is related to lattice distortion or molecular distortion.

Raman scattering is produced by coherent interactions between photoexcited electrons and phonons. Therefore, the number of photoexcited electrons greatly influences the Raman scattering coefficient. As a result, Raman scattering can be classified into non-resonant and resonant Raman scattering, as shown in figure 4.10. When the photon energy of the excitation is lower than the absorption bandgap ($E_g = E_c - E_v$), the scattering is defined as non-resonant Raman scattering. When the photon energy of the excitation is .greater than the absorption bandgap, the scattering is defined as resonant Stokes Raman scattering. When electrons are excited from molecular vibrational energy levels, the scattering is defined as resonant anti-Stokes Raman scattering. The Stokes and anti-Stokes types of Raman scattering can be observed simultaneously when a band-stop filter is used to block the excitation laser in front of the optical spectrum analyzer. In other words, the filter component in figure 4.1 must be a band-stop filter. In addition, coherent interactions (collisions) indicate that Raman scattering consists of polarized light waves which are related to the crystal or molecular structure. In figure 4.1, the linear polarizer (LP) in front of the filter can be used to analyze the polarization states of Raman scattering.

PL is strongly related to the optical bandgap and crystal structure of photoexcited materials. The difference between the absorption peak and PL peak energy values is defined as the Stokes shift. In many semiconductors, the Stokes shift and exciton binding energy (E_b) are almost the same. The E_g and E_b values of various inorganic semiconductors are listed in table 4.1. In Ga-based semiconductors, larger values of E_g correspond to larger values of E_b, which is related to the lattice constant. In a fixed lattice structure, bigger atoms result in a longer bond length and, therefore, a larger lattice constant. Conceptually, the E_g and E_b values of GaAs$_x$N$_{1-x}$ alloys can be varied from 4 meV and 1.42 eV to 28 meV and 3.40 eV as the x value increases from zero to one. In other words, the absorbance and PL spectra provide useful information that can be used to investigate the excitonic and crystal structural properties of various semiconductors.

The E_b values of semiconductors can also be obtained by analyzing the temperature-dependent PL spectra. In a bulk crystal, light is emitted by radiative excitons. When the binding energy between the electron–hole pairs (excitons) is lower than the thermal energy (K_BT), the excitons can thermally dissociate, thereby reducing the light emission intensity. The temperature-dependent light emission intensity can be expressed as equation (4.6), where K_BT is thermal energy, A is a temperature-independent constant, and I_0 is the PL intensity at $T = 0$. When E_b is far larger than K_BT, the emission intensity is insensitive to temperature. When E_b is very small, the PL intensity decreases from I_0 to $I_0/(1 + A)$ as the temperature increases from 0 K to infinity.

$$I(T) = \frac{I_0}{1 + Ae^{-E_b/K_BT}} \tag{4.6}$$

The PL width of semiconductors is mainly related to thermal broadening and inhomogeneous broadening. At $T = 0$, PL peak broadening is attributed to lattice disorder. Disordered lattices also result in a decrease in carrier relaxation time owing to the increased collision mechanisms. Therefore, larger lattice distortion results in a

Table 4.1. Stokes shift and E_b values of various inorganic semiconductors.

Inorganic semiconductor	E_g(eV)	E_b (meV)	References
Ge	0.66	2.7	[4]
Si	1.10	10	[4]
GaAs	1.42	4	[5]
GaN	3.40	28	[6]
CsPbBr$_3$	2.36	37	[7, 8]

broader PL width and a shorter relaxation time. In time-dependent PL decay curves, a two-constant exponential decay function has been widely used to compute the lifetimes of radiative excitons and non-radiative excitons. The rate equation for the time-dependent PL intensity can be written as equation (4.7a), where I is the time-dependent PL intensity, k is the relaxation coefficient, and the subscript R (NR) denotes radiative (non-radiative). The rate equation shows that the radiative and non-radiative relaxation paths are individual and separate. After solving the differential equation (equation 4.7a), the two-constant exponential decay function can be written as equation (4.7b), where τ_R ($=1/k_R$) is the lifetime of radiative excitons and τ_{NR} ($=1/k_{NR}$) is the lifetime of non-radiative excitons. In general, τ_{NR} is shorter than τ_R because of the driving electric field produced by the defect energy levels [9].

$$\frac{d(I_R + I_{NR})}{dt} = -k_R I_R - k_{NR} I_{NR} \tag{4.7a}$$

$$I_{PL}(t) = I_R e^{-t/\tau_R} + I_{NR} e^{-t/\tau_{NR}} \tag{4.7b}$$

4.4 Near-field optical spectrometry

When optical and optoelectronic components are fabricated on a flat substrate, light waves strongly interact with the structural materials. This behavior can be numerically simulated and designed using the finite-difference time-domain and finite-difference frequency-domain methods. The simulation methods are based on Maxwell's equations, which are described in equation (1.1a) of chapter 1. In general, near-field optical simulation is a powerful tool which can be used to design various optical waveguide-based components, such as power BSs, PBSs, polarization converters, filters, optical cavities, interferometers, subwavelength prisms, and optical limiters. However, it is extremely important to observe light–material interactions at the subwavelength scale and/or the nanoscale in order to confirm the simulation results. Figure 4.11 plots the optical configurations of two types of SNOMs [10, 11]. In general, the aperture-type and scattering-type SNOMs are based on shear-force atomic force microscopy (AFM) and tapping-mode AFM, respectively. In shear-force AFM, the tip–sample distance is fixed at about 2 nm, and the horizontal vibration amplitude of the tip ranges from 1 to 2 nm. In tapping-mode AFM, the average tip–sample distance is about 2 nm, and the vertical vibration amplitude of the tip is about 1 nm. In other words, environmental noise must be effectively reduced in order to suppress the influence of the tip–sample distance, which determines the collection efficiency of the near field.

In the aperture-type SNOM, a metal-coated tapered fiber is used as a nanoprobe (nanoexcitation) tip for bottom excitation (top excitation), as shown in figure 4.11(a). In general, the diameter of the smallest tip apex is about 100 nm when an Au/Pt-coated SiO_2 tapered fiber is used to collect light waves. In other words, the spatial resolution of the near-field images is about 100 nm. When the tip apex of the tapered fiber is smaller than 100 nm, transmission can be very low due to the large modal index

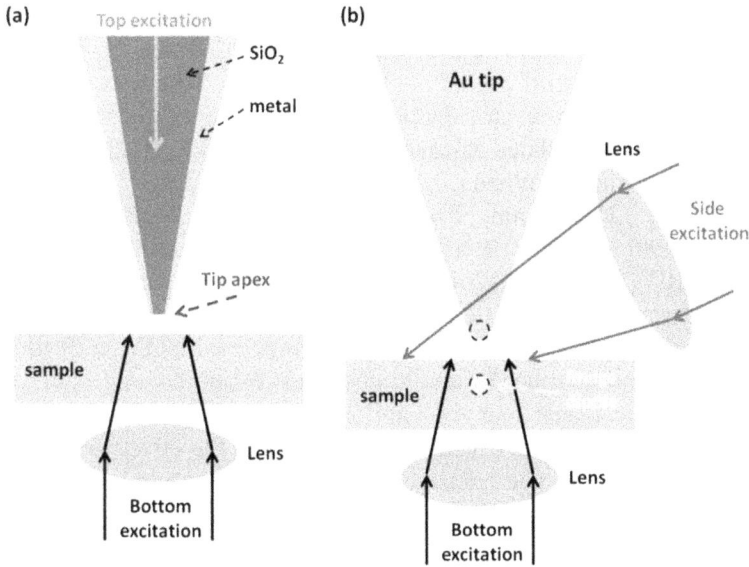

Figure 4.11. (a) Aperture-type SNOM. (b) Scattering-type SNOM.

mismatch between the tip apex and the optical waveguide-based components. The modal index of the waveguide-based components is lower than the refractive index of the guiding layer. The modal index of the metal-coated fiber greatly increases as the diameter of the tip apex decreases because surface plasmon waves (SPWs) dominate the optical mode. When the radius of the fiber tip apex is smaller than the penetration depth of the SPWs, the modal index is greatly increased owing to the strong field localization effect. In a dielectric medium, the penetration depth of the SPWs can be computed using equations (3.10d) and (3.10g). When the relative permittivity of SiO_2 (Au) is 2.1025 ($-115 + i11.3$) at $\lambda_0 = 1550$ nm, the calculated penetration depth in SiO_2 is about 1246 nm, which is far larger than the radius of the fiber tip apex. In other words, strong localized fields result in a large modal index at the end of the metal-coated tapered fiber, thereby causing large reflectivity at the air/fiber interface. Conceptually, the field localization effect can be reduced if a high-index Si fiber is used to replace the SiO_2 fiber. When the relative permittivity of Si (Au) is 12.089 ($-115 + i11.3$) at $\lambda_0 = 1550$ nm, the calculated penetration depth in Si is about 207 nm. In other words, the diameter of a metal-coated Si tapered fiber end can be made smaller than 100 nm while retaining acceptable transmission.

In scattering-type SNOM, the apex of the Au tip can be approximated as a sphere with a subwavelength radius, which can be viewed as an electric dipole under light illumination. The optically excited electric dipole induces an image dipole in the sample, which results in effective polarizability. The polarizability of an electric dipole can be written as equation (4.8a), where r is the radius of the tip apex and ε_t is the relative permittivity of the tip. The polarizability of the image dipole is proportional to the electric dipole, which can be written as equation (4.8b), where ε_s is the relative permittivity of the sample. When the electric field is perpendicular to the sample surface, the effective polarizability (α_{eff}) is related to the tip–sample distance (z), which can be written as equation (4.8c), where $2r + 2z$ equals the distance between the central points of the two dipoles. Then, α_{eff} can be used to evaluate the light absorption and light scattering from the tip–sample interaction region, which can be formulated as equations (4.8d) and (4.8e) [12], where k ($=\lambda_0/2\pi$) is the propagation constant. When r, ε_t, ε_s, z, and λ_0 are 20 nm, $-115 + i11.3$, $-115 + i11.3$, 2, and 1550 nm, respectively, the calculated α_{eff} equals $(26.1908 + i0.0900) \times (20 \times 10^{-9}m)^3$. The calculated values of C_{abs} and C_{sca} are then 2.8848×10^{-18} m^2 and 4.5261×10^{-19} m^2, respectively, which shows that the absorption is larger than the scattering in the tip–sample interaction region when the tip and sample are both made from Au. In scattering-type SNOM, a tilted laser beam is focused on the sample surface by an objective lens. When the long axis (r_a) and short axis (r_b) of the elliptical laser spot are 4 and 2 μm on the sample surface, respectively, the excitation area is 8π μm^2, which is far larger than the effective area of the tip–sample interaction region. The effective area of the tip–sample interaction region is related to the radius of the tip apex because the electric field is concentrated near the Au tip apex. Therefore, the cross-sectional area of the tip apex can be viewed as the effective area of the tip–sample interaction region, which equals 400π nm^2. The ratio of the effective area (πr^2) to the excitation area ($\pi r_a r_b$) can be used to evaluate the tip–sample scattering efficiency, which is about 5×10^{-5}. In other words, the highest scattering power in the tip–sample interaction region is 50 μW when the excitation power is 1 W. The intensity of the elliptical laser beam is about 3.97 MW cm^{-2} when the power and area are 1 W and 8π μm^2, respectively. However, the damage threshold of Au is about 200 W cm^{-2}. In other words, the excitation power must be lower than 50 μW, which shows that the highest scattering power that can be produced by the tip–sample interaction region is about 2.5 nW. Fortunately, a lock-in amplifier and a heterodyne interferometer can be used to effectively amplify the weak scattering signal produced by the tip–sample interaction region [13].

$$\alpha_t = 4\pi r^3[(\varepsilon_t - 1)/(\varepsilon_t + 2)] \tag{4.8a}$$

$$\alpha_i = \beta\alpha_t = [(\varepsilon_s - 1)/(\varepsilon_s + 1)]\alpha_t \tag{4.8b}$$

$$\alpha_{\text{eff}} = \frac{\alpha_t + \alpha_i}{1 - \alpha_i/[2\pi(2r + 2z)^3]} \tag{4.8c}$$

$$C_{\text{abs}} = k\,\text{Im}[\alpha_{\text{eff}}] \tag{4.8d}$$

$$C_{\text{sca}} = k^4\,|\alpha_{\text{eff}}|^2/6\pi \tag{4.8e}$$

4.5 Surface-enhanced Raman scattering effects

To increase weak Raman scattering signals, the near fields of SPWs can be used to enhance the local electric field near the target molecules. Rhodamine 6G (R6G) molecules are widely used to evaluate surface-enhanced Raman scattering (SERS)

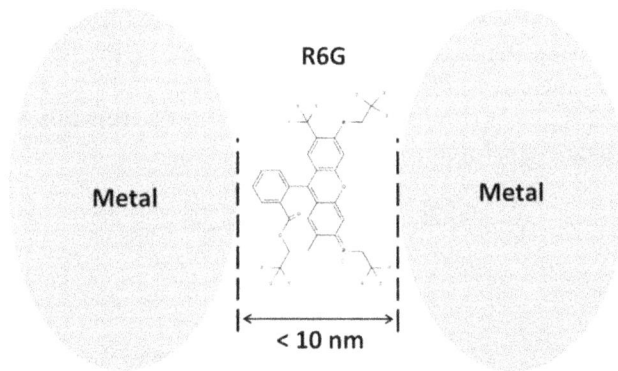

Figure 4.12. An R6G molecule located in between two Au nanoparticles.

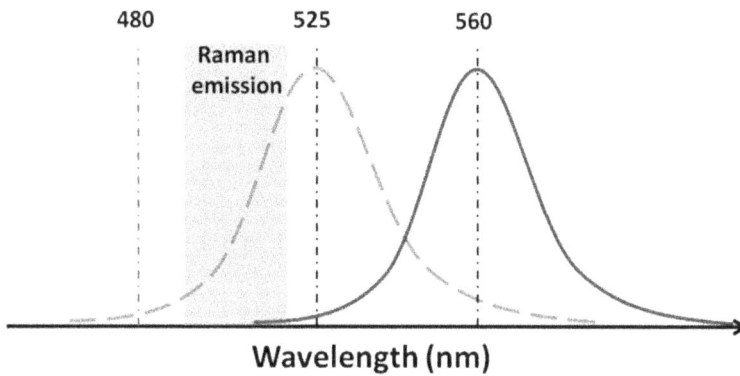

Figure 4.13. Absorption peak and photoluminescence peak of R6G molecules.

effects, as shown in figure 4.12. The Raman shift (Δ_{RS}) can be computed using equation (4.9a), where λ_L and λ_R are the excitation wavelength and the Raman emission wavelength, respectively. The relationship between Δ_{RS} and λ_R can be expressed as equation (4.9b). It should be noted that the unit of Raman shift is cm^{-1}. The absorption and PL peak wavelengths of R6G molecules are 525 and 560 nm, respectively, as shown in figure 4.13. The excitation wavelength required to effectively excite R6G molecules is 525 nm. The surface-enhanced optical effects of metallic nanoparticles have been used to detect R6G molecules at low solution concentrations. This detection was achieved using Raman scattering at an excitation wavelength of 532 nm. However, the Raman shift of R6G molecules ranges from 1100 to 1700 cm^{-1}, which corresponds to a wavelength range from 565.1 to 588.2 nm when the excitation wavelength is 532 nm. In other words, the Raman emission and PL fall within the same wavelength range when the excitation wavelength matches the excitonic transition peak wavelength of R6G molecules near the metal surface, which is defined as surface-enhanced resonant Raman scattering.

Conceptually, the Raman scattering and PL of molecules are influenced by each other, which increases the difficulty of understanding SERS effects. To avoid the influence of PL on SERS effects, the excitation wavelength should be far from the absorption peak wavelength of the target molecules. When the excitation wavelength is 480 nm, the range of Raman shifts is from 1100 to 1700 cm^{-1}, which corresponds to the range of Raman emission wavelengths from 506.8 nm to 522.6, as depicted in the blue box of figure 4.13.

$$\Delta_{RS} = \left(\frac{1}{\lambda_L} - \frac{1}{\lambda_R} \right) = \left(\frac{\lambda_R - \lambda_L}{\lambda_L \lambda_R} \right) \tag{4.9a}$$

$$\lambda_R = \frac{\lambda_L}{1 - (\Delta_{RS})\lambda_L} \tag{4.9b}$$

When the excitation wavelength is 480 nm, the surface plasmon resonances (SPRs) of Ag or Al nanoparticles can be used to increase the strength of local electric fields while retaining relatively low ohmic loss in the metal region. In other words, the electric fields caused by excitation at a wavelength of 480 nm can be made to mainly accumulate near the metal surfaces, thereby increasing the Raman scattering of the R6G molecules while avoiding the influence of PL on the SERS spectrum. However, surface oxidation is the main drawback of Ag and Al nanoparticles, as it decreases the strength of the electric fields that accumulate on the target molecules.

To avoid the influence of PL and thermal effects on the SERS spectrum of the target molecules, the excitation wavelength should be longer than 800 nm when the SPRs of Au nanoparticles are used. The use of Au nanoparticles leads to better chemical stability during measurements. The use of near-infrared excitation can result in low ohmic loss in the Au nanoparticles, thereby reducing excitation-induced heating effects on the SERS spectrum. It should be noted that the near-infrared excitation is not effectively absorbed by the Au nanoparticles and target molecules, which is known as surface-enhanced non-resonant Raman scattering.

Bibliography

[1] Zhu H and Blackborow P 2019 Laser-driven light sources for nanometrology applications *J. Microelectron. Manuf.* **2** 19020104
[2] Polyanskiy M N 2024 Refractiveindex.info database of optical constants *Sci. Data* **11** 94
[3] Thakur D and Chang S H 2024 Material properties and optoelectronic applications of lead halide perovskite thin films *Synth. Met.* **301** 117535
[4] Macfarlane G G, Mclean T P, Quarrington J E and Roberts V 1959 Exciton and phonon effects in the absorption spectra of germanium and silicon *J. Phys. Chem. Solids* **8** 388–92
[5] Sukumar B and Navaneethakrishnan K 1990 Effect of the dielectric function and pressure on the binding energies of excitons in GaAs and GaAs/Ga$_{1-x}$Al$_x$As superlattices *Solid State Commun.* **76** 561–4
[6] Reimann K, Steube M, Frohlich D and Clarke S J 1998 Exciton binding energies and bandgap bulk crystals *J. Cryst. Growth* **189** 652–5
[7] Yuan Y, Chen M, Yang S, Shen X, Liu Y and Cao D 2020 Exciton recombination mechanisms in solution single crystalline CsPbBr$_3$ perovskite *J. Lumin.* **226** 117471

[8] Ezzeldien M, Al-Qaisi S, Alrowaili Z A, Alzaid M, Maskar E, Es-Smairi A, Vu T V and Rai D P 2021 Electronic and optical properties of bulk and surface of $CsPbBr_3$ inorganic halide perovskite a first principles DFT 1/2 approach *Sci. Rep.* **11** 20622

[9] Thakur D, Ke Q B, Chaing S-E, Tseng T-H, Cai K-B, Yuan C-T, Wang J-S and Chang S H 2022 Stable and efficient soft perovskite crystalline film based solar cells prepared with a facile encapsulation method *Nanoscale* **14** 17625–32

[10] Novotny L, Pohl D W and Regli P 1995 Near-field, far-field and imaging properties of the 2D aperature SNOM *Ultramicroscopy* **57** 180–8

[11] Kelmann F and Hillenbrand R 2004 Near-field microscopy by elastic light scattering from a tip *Phil. Trans. R. Soc. Lond.* A **362** 787–805

[12] Knoll B and Keilmann F 1999 Near-field probing of vibrational absorption for chemical microscopy *Nature* **399** 134–6

[13] Cheng T-Y, Wang H-H, Chang S H, Chu J-Y, Lee J-H, Wang Y-L and Wang J-K 2013 Revealing local, enhanced optical field characteristics of Au nanoparticle arrays with 10 nm gap using scattering-type scanning near-filed optical microscopy *Phys. Chem. Chem. Phys.* **15** 4275–82

IOP Publishing

Light–Material Interactions and Applications in Optoelectronic Devices

Anjali Chandel and Sheng Hsiung Chang

Chapter 5

Optical waveguide-based sensors

This chapter describes dielectric waveguides, plasmonic waveguides, and their applications. The propagation constants and modal field distributions of waveguides are obtained using transcendental equations with a graphical method and are then used to discuss modal dispersion and polarization dispersion. The propagation characteristics and field distributions of the waveguide modes are analytically resolved and conceptually discussed in order to understand the design rules of various waveguide-based sensors. To predict the sensitivities of waveguide-based sensors, the transfer matrix method is used to compute changes in the transmittance and/or reflectance spectra. The optical devices discussed here are distributed Bragg reflector-based waveguide cavities, lossy mode resonance-based sensors, and dielectric/plasmonic/dielectric waveguide-based sensors, which can be used as examples when starting to learn the design concepts of optical waveguide-based sensors.

5.1 Principle of optical waveguides

In an optical guiding layer, collimated beams can be guided in the high-index region if the requirements for total internal reflection are satisfied at the top and bottom interfaces, as shown in figure 5.1(a). The incident angle (θ_i) must be larger than the critical angle (θ_c) of the dielectric/air interface. When n_0 and n_{core} are 1.0 and 1.5, respectively, θ_c is computed to be about 41.8°. In other words, the θ_i value of the collimated beam must be larger than 41.8°. It should be noted that the t_g value of the optical guiding layer is far larger than the wavelength of the EM wave. When t_g (the thickness of the guiding layer) is not much larger than the wavelength of the EM wave, wave interference in the guiding layer results in a fixed modal profile and an effective propagation direction, as shown in figure 5.1(b), which is defined as a slab waveguide.

doi:10.1088/978-0-7503-6099-9ch5

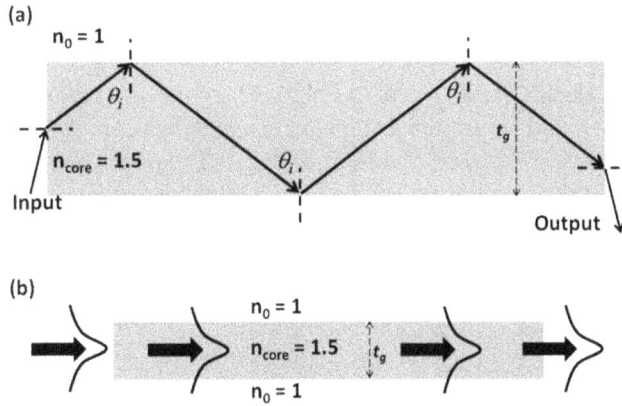

Figure 5.1. (a) The path of a collimated beam in an optical guiding layer. (b) The modal profile of a symmetric slab waveguide.

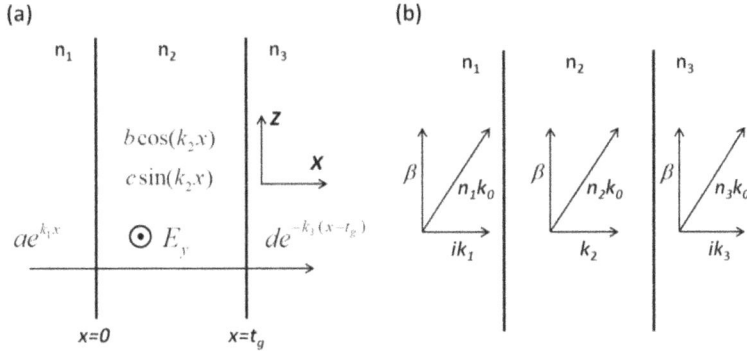

Figure 5.2. (a) The modal profile of the transverse electric mode in a slab waveguide. (b) The β–k_0 relations.

In a slab waveguide, the modal profile of the electric field is the combination of a cosine function and a sine function in the core region, and the electric fields take the form of exponential decay curves in the left-hand (n_1) and right-hand (n_3) regions, as shown in figure 5.2(a), where a, b, c, and d are coefficients and n_1, n_2, and n_3 are refractive index values. The relation between the propagation constant (β) of the waveguide mode and the operating wavelength (λ_0) is plotted in figure 5.2(b), which can be used to compute the x-directed propagation constants k_1, k_2, and k_3. The k_1, k_2, and k_3 values determine the modal profiles of the slab waveguide. The geometrical relations can be written as equations (5.1a), (5.1b), and (5.1c), where k_0 equals $2\pi/\lambda_0$.

$$k_1 = \sqrt{\beta^2 - n_1^2 k_0^2} \tag{5.1a}$$

$$\beta = \sqrt{n_2^2 k_0^2 - k_2^2} \tag{5.1b}$$

$$k_3 = \sqrt{\beta^2 - n_3^2 k_0^2} \qquad (5.1c)$$

The y-directed electric field (E_y) of the slab waveguide can be written as equation (5.2a), where k_1, k_2, and k_3 are the x-directed propagation constants for the left-hand ($x \leqslant 0$), core ($0 \leqslant x \leqslant t_g$), and right-hand ($t_g \leqslant x$) regions, respectively. The four coefficients a, b, c, and d can be determined by using the boundary conditions at $x = 0$ and $x = t_g$. The tangential electric fields (E_y) are continuous across the two interfaces. At $x = 0$, $a = b$ is obtained. At $x = t_g$, $b\cos(k_2 t_g) + c\sin(k_2 t_g) = d$ is obtained. The tangential magnetic fields (H_z) can be obtained using the source-free form of Faraday's law, which can be written as equation (5.2b), where μ is permeability. The tangential magnetic fields (H_z) are continuous across the two interfaces. At $x = 0$, $ak_1 = ck_2$ is obtained. At $x = t_g$, $-bk_2 \sin(k_2 t_g) + ck_2 \cos(k_2 t_g) = -dk_3$ is obtained. After the four relations have been solved, the transcendental equation for the transverse electric (TE) modes can be written as equation (5.2c). Equations (5.1a), (5.1b), (5.1c), and (5.2c) can be used to solve the propagation constant (β) values using the graphical method when t_g, n_1, n_2, n_3, and k_0 ($=2\pi/\lambda_0$) are known. In the graphical method, $\tan(k_2 t_g)$ and $(k_1 + k_3)/(k_2 - k_1 k_3/k_2)$ are denoted by Y_1 and Y_2, respectively. The intersections of the Y_1 and Y_2 curves give the solutions, which can be used to determine the values of β, k_1, k_2, and k_3.

$$E_y(x) = \begin{cases} ae^{k_1 x}, \; x \leqslant 0 \\ b\cos(k_2 x) + c\sin(k_2 x), \; 0 \leqslant x \leqslant t_g \\ de^{-k_3(x - t_g)}, \; x \geqslant t_g \end{cases} \qquad (5.2a)$$

$$\nabla \times \vec{E} = -\partial(\mu \vec{H})/\partial t \qquad (5.2b)$$

$$\tan(k_2 t_g) = \frac{k_1 + k_3}{k_2 - k_1 k_3/(k_2)} \qquad (5.2c)$$

Let us practice solving the propagation constants, modal index values, and modal profiles of a slab waveguide when the required parameters are given as follows. The Y_1 and Y_2 curves of the transcendental equation are plotted in figure 5.3 for t_g, n_1, n_2, n_3, and λ_0 values of 5000, 1.0, 1.5, 1.0, and 1550 nm, respectively. The eight crossing points (β values) are 6.05×10^6 m^{-1}, 5.94×10^6 m^{-1}, 5.82×10^6 m^{-1}, 5.63×10^6 m^{-1}, 5.36×10^6 m^{-1}, 5.02×10^6 m^{-1}, 4.61×10^6 m^{-1}, and 4.12×10^6 m^{-1}. These can be used to compute the modal index values of the slab waveguide using the simple relation: $n_{\text{modal}} = \beta/k_0$. The calculated modal index values of the TE modes are 1.493, 1.473, 1.444, 1.389, 1.322, 1.238, 1.136, and 1.016. The largest modal index value corresponds to the fundamental mode. When the β value of the fundamental mode is 6.06×10^6 m^{-1}, the calculated k_1 (k_3) and k_2 values are 4.49×10^6 m^{-1} and 0.60×10^6 m^{-1}, respectively, which can be used to solve the relationship between coefficients a, b, c, and d. When a equals one, b, c, and d are one, 7.449, and one, respectively. The black solid line and red dashed line are the modal profiles of the fundamental and first higher-order modes in the core region, respectively, as shown

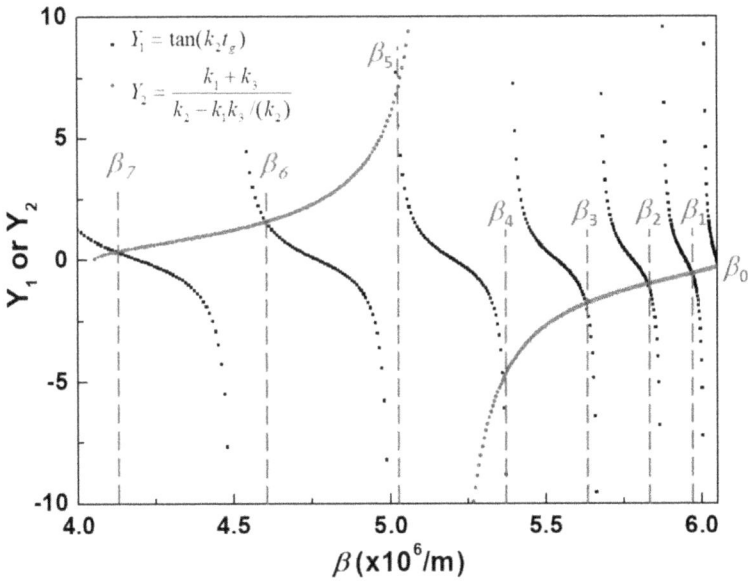

Figure 5.3. Graphical method for solving the β values of the transverse electric (TE) mode.

$Y_1 = \tan(k_2 t_g)$

$Y_2 = \dfrac{k_1 + k_3}{k_2 - k_1 k_3 / (k_2)}$

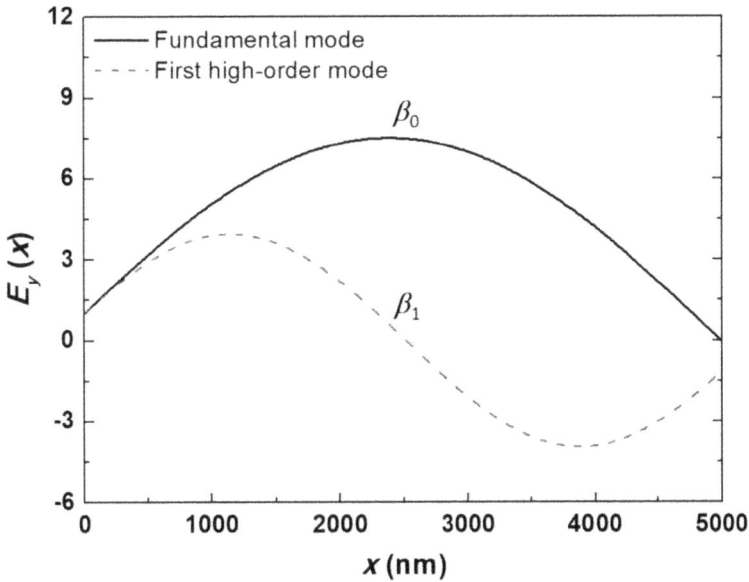

Figure 5.4. Electric field distributions in the core region of a symmetric slab waveguide.

in figure 5.4. In the n_1 (n_3) region, the evanescent tail of the fundamental mode equals $1/k_1$ ($1/k_3$), which is about 223 nm. The evanescent tails are far smaller than the t_g value (5000 nm), which indicates that the fundamental mode is mainly confined to the core region. The discrete β values indicate that the formation of the waveguide modes must satisfy the resonance condition in the x-direction and the conditions for total internal reflection at the two interfaces ($x = 0$ and $x = t_g$). According to equations (5.1a), (5.1b), and (5.1c), larger β values correspond to smaller x-directed propagation constants (k_1, k_2, and k_3), which indicates that the incident angle from the n_2 region to the n_1 (n_3) region is larger, and the evanescent tail (penetration depth) in the n_1 (n_3) region is shorter.

In the transverse magnetic (TM) modes, the magnetic field is polarized in the y-direction, which can be written as equations (5.3a). The four coefficients a, b, c, and d can be determined by solving the boundary conditions at $x = 0$ and $x = t_g$. The tangential magnetic fields (H_y) are continuous across the two interfaces. At $x = 0$, $a = b$ is obtained. At $x = t_g$, $b \cos(k_2 t_g) + c \sin(k_2 t_g) = d$ is obtained. The tangential electric fields (E_z) can be obtained using the source-free form of Ampère's law, which can be written as equation (5.3b). The tangential electric fields (E_z) are continuous across the two interfaces. At $x = 0$, $ak_1/(\omega \varepsilon_1) = ck_2/(\omega \varepsilon_2)$ is obtained. At $x = t_g$, $[bk_2^2/(\varepsilon_2)]\sin(k_2 t_g) - [ck_2^2/(\varepsilon_2)]\cos(k_2 t_g) = dk_3/(\varepsilon_3)$ is obtained. After the four relations have been solved, the transcendental equation of the TM modes can be written as equation (5.3c). A comparison of equations (5.2c) and (5.3c) shows that the x-directed propagation constant of the TM mode is smaller than that of the TE mode. In other words, the evanescent tail of the TM mode is longer than that of the TE mode. When t_g, n_1, n_2, n_3, and λ_0 are 5000 nm, 1.0, 1.5, 1.0, and 1550 nm, respectively, the computed modal index values of the eight TM modes are 1.493, 1.470, 1.431, 1.379, 1.305, 1.215, 1.106, and 1.009, respectively.

$$H_y(x) = \begin{cases} ae^{k_1 x}, \, x \leqslant 0 \\ b \cos(k_2 x) + c \sin(k_2 x), \, 0 \leqslant x \leqslant t_g \\ de^{-k_3(x - t_g)}, \, x \geqslant t_g \end{cases} \tag{5.3a}$$

$$\nabla \times \vec{H} = \partial \vec{D}/\partial t \tag{5.3b}$$

$$\tan(k_2 t_g) = \frac{k_1 k_2(n_2^2/n_1^2) + k_3 k_2(n_2^2/n_3^2)}{k_2^2 - k_1 k_3[n_2^4/(n_1^2 n_3^2)]} \tag{5.3c}$$

The modal index values of the TE and TM modes are listed in table 5.1. Δn is defined as the difference between the modal indexes of the TE and TM modes. In the fundamental mode, Δn is almost zero, which indicates that the polarization dispersion can be ignored. The higher-order mode has a larger Δn value, which results in stronger polarization dispersion. However, the highest-order mode (β_7) does not have the largest polarization dispersion, which can be explained by the long evanescent tails. Equations (5.1a) and (5.1c) show that k_1 (k_3) is very small when the value of β is close to $n_1 k_0$ ($n_3 k_0$), which results in a long evanescent tail in the n_1 (n_3) region. Therefore, the modal index of the highest-order mode of a symmetric slab waveguide is close to the refractive index in the n_1 (n_3) region, thus leading to lower polarization dispersion.

In the TE modes, the β_0 values of the symmetric slab waveguide are 6.06×10^6 m^{-1}, $6.04 \times 10^6 \, m^{-1}$, $6.01 \times 10^6 \, m^{-1}$, $5.94 \times 10^6 \, m^{-1}$, and $5.69 \times 10^6 \, m^{-1}$ when the

Table 5.1. Modal index values of TE and TM modes of a symmetric slab waveguide.

Mode	TE	TM	Δn
$n_1 = n_3 = 1.0$, $n_2 = 1.5$, $t_g = 5000$ nm, $\lambda_0 = 1550$ nm			
β_0	1.493	1.493	0.000
β_1	1.473	1.470	0.003
β_2	1.444	1.437	0.007
β_3	1.389	1.377	0.012
β_4	1.322	1.305	0.017
β_5	1.238	1.215	0.023
β_6	1.136	1.106	0.030
β_7	1.016	1.009	0.007

t_g values are 5000, 4000, 3000, 2000, and 1000 nm, respectively. The t_g-dependent β_0 value indicates that lower values of t_g result in smaller incident angles (θ_i) at the n_2/n_1 and n_2/n_3 interfaces, thereby resulting in longer evanescent tails in the n_1 and n_3 regions.

When t_g, n_1, n_2, n_3, and λ_0 are 5000 nm, 1.0, 1.5, 1.45, and 1550 nm, respectively, there are three crossing points (two crossing points) for the TE (TM) mode. The β values of the TE (TM) mode are 6.05×10^6 m^{-1} (6.05×10^6 m^{-1}), 5.99×10^6 m^{-1} (5.98×10^6 m^{-1}), and 5.88×10^6 m^{-1} (none). In other words, asymmetric slab waveguides have a cutoff condition because the requirement for total internal reflection cannot be satisfied at the n_2/n_3 interface. The E_y distribution of the fundamental mode in the core region is plotted in figure 5.5, which shows that the peak position of the asymmetric modal profile is close to the n_2/n_3 interface. The n_1 and n_3 regions can be viewed as the air environment and the substrate, respectively.

As t_g decreases, the incident angle of the fundamental mode approaches the critical angle (θ_c). When the incident angle (θ_i) at the n_2/n_1 and n_2/n_3 interfaces equals the critical angle (θ_c) at the n_2/n_3 interface, the thickness of the guiding layer (t_g) required to support the fundamental mode can be assumed to be π/k_2, as shown in figure 5.6. When n_1, n_2, and n_3 are 1.0, 1.5, and 1.45, respectively, the critical angle at the n_2/n_3 interface is about 75.16°; this can be used to determine the k_2 value via the geometric relation $\cos\theta_c = k_2/(n_2k_0)$. The calculated k_2 value is 1.56×10^6 m^{-1} when λ_0 is 1550 nm, which can be used to compute the critical t_g. The critical t_g value of the asymmetric slab waveguide is 2017 nm when t_g is assumed to be π/k_2. When t_g is smaller than π/k_2, the incident angle (θ_i) is smaller than the critical angle (θ_c) at the n_2/n_3 interface, thereby resulting in a lossy mode. In other words, a guided mode cannot form in an asymmetric slab waveguide when the thickness of the guiding layer is less than the critical value of t_g.

In optical integrated circuits, the modal profiles of optical waveguides are two-dimensionally bounded. The field distributions and modal index values of the

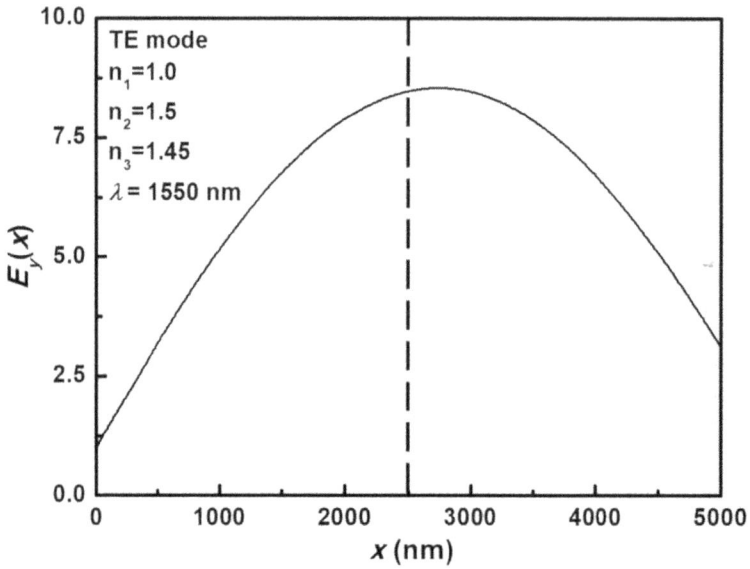

Figure 5.5. Electric field distribution of the fundamental mode in the core region of an asymmetric slab waveguide.

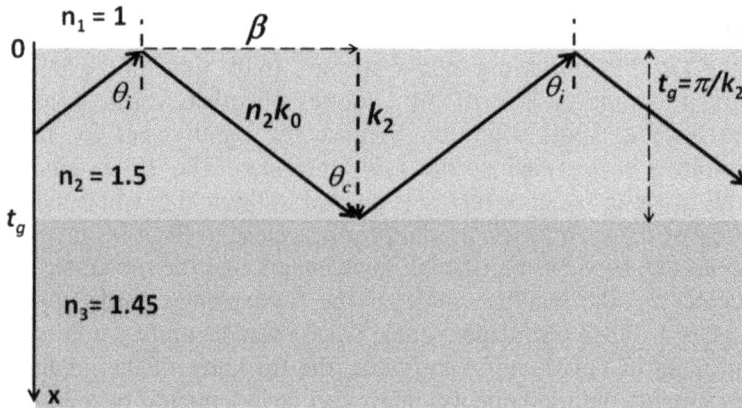

Figure 5.6. Determination of the critical t_g value of an asymmetric slab waveguide using the geometric relationships between $n_2 k_0$, β, and k_2.

waveguide mode can be solved using the effective medium method which is described as follows. Figure 5.7 plots a cross-sectional view of a dielectric strip waveguide, where the width and height of the dielectric strip are b and a, respectively. The EM waves are guided in the negative z-direction when the polarizations of E and H are in the positive y and positive x-directions, respectively. The value of n_{core} (n_s) is larger than that of n_s (n_{air}). In the horizontal direction (x-direction), the two boundaries are separated by a distance b. Therefore, E_y is defined as the TE mode. The transcendental equation of the TE mode can be written

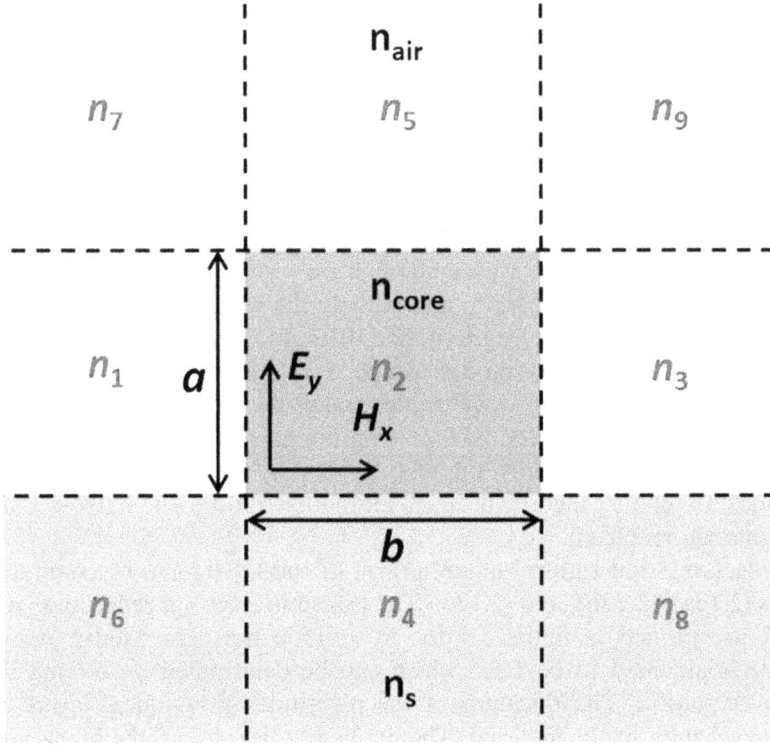

Figure 5.7. Cross-sectional structure of a dielectric strip waveguide.

as equation (5.4a) [1]. The solutions of β_{x2}, k_{x2}, k_1, and k_3 in the TE mode determine the field distribution in the horizontal direction and the effective index. The effective index of the TE mode can be computed using the simple relation: $n_{\text{eff}} = \beta_{x2}/k_0$. In other words, the $n_1/n_2/n_3$ structure can be viewed as a slab waveguide. The $n_6/n_4/n_8$ ($n_7/n_5/n_9$) structure can also be viewed as a slab waveguide. The effective index of the $n_6/n_4/n_8$ ($n_7/n_5/n_9$) structure is n_s (n_{air}). In the vertical direction (y-direction), the two boundaries are separated by a distance a. Therefore, H_x is defined as the TM mode. The transcendental equation for the TM mode can be written as equation (5.4b), where n_{eff} is the computed result obtained from equation (5.4a) [1]. The solutions of β_{y2}, k_{y2}, k_4, and k_5 in the TM mode determine the field distribution in the vertical direction and the modal index of the waveguide mode. The modal index of the waveguide mode can be calculated using the simple relation: $n_m = \beta_{y2}/k_0$.

$$\tan(k_{x2}b) = \frac{k_1 + k_3}{k_{x2} - k_1 k_3/(k_{x2})} \tag{5.4a}$$

$$\tan(k_{y2}a) = \frac{k_4 k_{y2}(n_{\text{eff}}^2/n_s^2) + k_5 k_{y2}(n_{\text{eff}}^2/n_{\text{air}}^2)}{k_{y2}^2 - k_4 k_5[n_{\text{eff}}^4/(n_s^2 n_{\text{air}}^2)]} \tag{5.4b}$$

5.2 Waveguide-based sensors

Optical waveguide-based sensors are based on cavity effects or lossy mode resonant effects, as shown in figure 5.8. The two distributed Bragg reflectors (DBRs) shown in figure 5.8(a) can be used as the mirrors of a waveguide cavity. L_c is the cavity length, which determines the free spectral range (FSR) between two adjacent longitudinal modes. The FSR can be computed using the simple relation: $\Delta\lambda_{FSR} \sim \lambda_0^2/2n_g L_c$, where n_g is the modal index of the waveguide mode. When the FSR is larger than the bandwidth of the DBRs, the waveguide can only support one cavity mode. In a single-longitudinal-mode cavity, L_c must satisfy the relation $L_c = \lambda_0/(2n_g)$. λ_0 can be used as the operating wavelength of the DBRs. In figure 5.8(b), the use of a high-index material supports lossy modes, which results in valleys in the transmission spectrum when the operating wavelengths match the resonant wavelengths of the lossy modes. In the lossy modes, EM waves propagate from the high-index material to the substrate. In other words, the lossy modes are radiative modes [2]. Thicker, higher-index materials generally produce more lossy modes and narrower valleys in the transmission spectrum.

The reflectance and transmittance spectra of the DBRs can be computed using equations (2.13a), (2.13b), and (2.13c). The transmittance and reflectance spectra of the DBR are plotted in figure 5.9 for $\lambda_0 = 1550$ nm. The modal index of the waveguide is assumed to be 1.55, which can be determined by solving the transcendental equation. The thickness of the high-index (low-index) layers equals a quarter wavelength in the material. The designed thickness of the high-index (low-index) layers of the DBR is 250 nm (387.5 nm). When the number of high-index/low-index pairs is six (three), the largest reflectance value of the DBR is 97.94% (74.91%) at the design wavelength. The larger number of pairs (N) in the DBR results in higher (lower) reflectance (transmittance), which indicates that the quality factor of

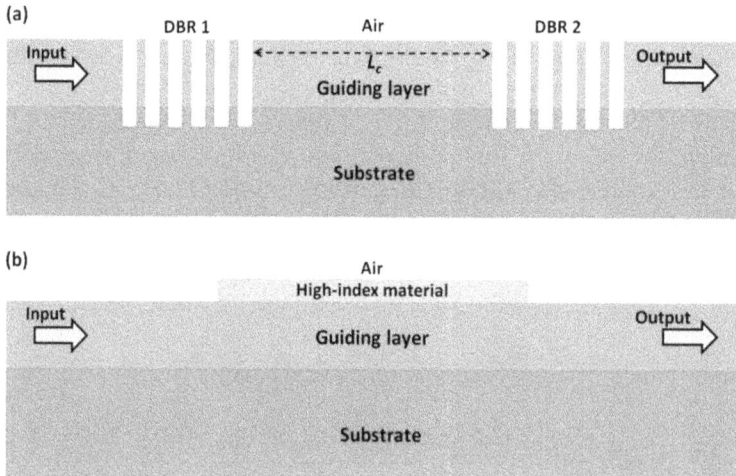

Figure 5.8. (a) Distributed Bragg reflector-based waveguide cavity. (b) An asymmetric waveguide with a lossy mode resonant structure.

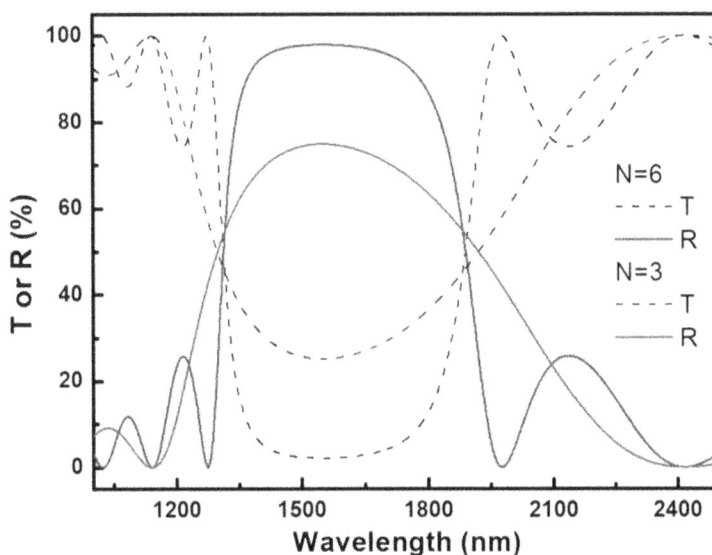

Figure 5.9. Transmittance and reflectance spectra of a DBR for six pairs and three pairs.

the waveguide-based cavity mode can be manipulated by varying the N value. The 3 dB bandwidth of the six-pair DBR is about 578 nm. When half of the 3 dB bandwidth is designed to equal $\Delta \lambda_{FSR}$, the value of L_c can be computed using the relation $\Delta \lambda_{FSR} \sim \lambda_0^2 / 2 n_g L_c$. The calculated value of L_c is about 2681 nm when λ_0 and n_g are 1550 nm and 1.55, respectively. It is predicted that three longitudinal modes can form in the waveguide-based cavity when the L_c value is 2681 nm.

Figure 5.10 plots the transmittance spectra of a waveguide-based cavity for different L_c values. When the L_c value is 2681 nm, the three longitudinal modes occur at wavelengths of 1360, 1523, and 1740 nm, respectively. The $\Delta \lambda_{FSR}$ between λ_1 (λ_2) and λ_2 (λ_3) is 163 nm (217 nm). The two $\Delta \lambda_{FSR}$ values are both smaller than the predicted value of 289 nm, which indicates that the effective cavity length is longer than L_c. In other words, the penetration depths in the DBRs contribute to the effective cavity length. When the L_c value is reduced to 2000 nm, two longitudinal modes occur at wavelengths of 1438 and 1681 nm, respectively. The $\Delta \lambda_{FSR}$ value of the two modes is 247 nm, which is larger than the two $\Delta \lambda_{FSR}$ values (163 and 217 nm) observed when the L_c value is 2681 nm.

Let us discuss the origins of the different $\Delta \lambda_{FSR}$ values by comparing the features of these modes, which are listed in table 5.2. $\Delta \lambda_c$ is defined as the relation $\Delta \lambda_c = \lambda_l - 1550$ nm, where λ_l is the peak wavelength of the longitudinal mode and 1550 nm is the design wavelength of the DBRs. The quality factor (Q) is defined as the relation: $Q = \Delta \lambda / \lambda_l$, where $\Delta \lambda$ is the peak width. Q is a metric that describes energy accumulation in a cavity. Larger values of $\Delta \lambda$ correspond to larger absolute values of $\Delta \lambda_c$. Larger values of $\Delta \lambda$ occur at lower values of Q, which can be explained by the lower reflectance of the DBR near the edge wavelengths of the reflection peak (see figure 5.9). In other words, the penetration depth in the DBR is longer when the

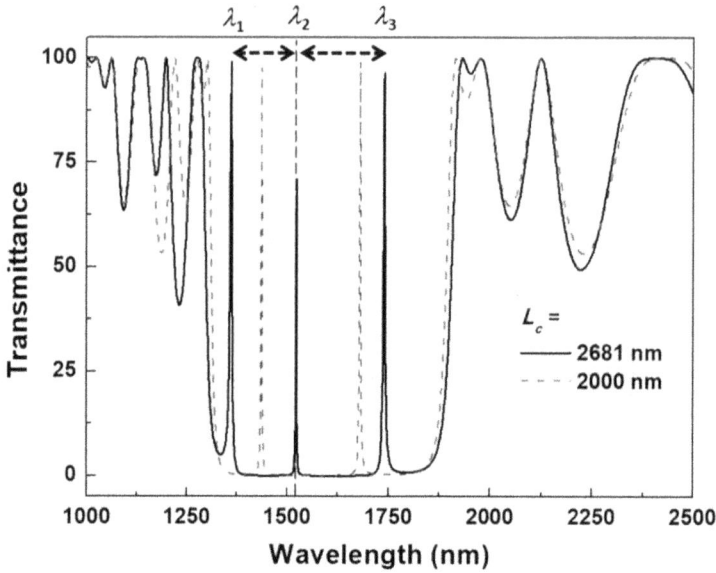

Figure 5.10. Transmittance spectra of a waveguide-based cavity with two DBRs.

Table 5.2. Peak wavelength (λ_l), $\Delta\lambda_c$, peak width ($\Delta\lambda$), and quality factor (Q) values of the five modes in figure 5.10.

λ_l (nm)	1360	1438	1523	1681	1740
$\Delta\lambda_c$ (nm)	−190	−112	−27	131	190
$\Delta\lambda$ (nm)	~4.7	~2.3	~2.0	~3.2	~4.5
Q	~289	~625	~761	~525	~386

peak wavelength of the longitudinal mode is closer to the edges of the reflection peak, which results in a longer effective cavity length, thereby reducing $\Delta\lambda_{\mathrm{FSR}}$.

When the L_c value is 800 nm, the waveguide-based cavity only supports one longitudinal mode, as shown in figure 5.11. The peak wavelength of this mode is related to the modal index (n_g) of the waveguide. In general, the modal index of the waveguide is proportional to the refractive index of the guiding layer. In other words, a change in the refractive index of the guiding layer can be detected by analyzing the properties of the longitudinal mode. The peak wavelength of the longitudinal mode linearly increases from 1586 to 1595 nm as the modal index increases from 1.54 to 1.56. When the refractive index of the guiding layer is related to temperature, the waveguide-based cavity can be used as a temperature sensor [3]. The temperature-dependent refractive index of dielectric materials can be expressed as equation (5.5a), where n_0 is the refractive index at $T = 0$ and dn/dT is the thermal coefficient of the refractive index. The dn/dT value of glass fiber is about 1×10^{-5} K^{-1} [4]. When the temperature is increased from 300 K to 1300 K, the increase in the

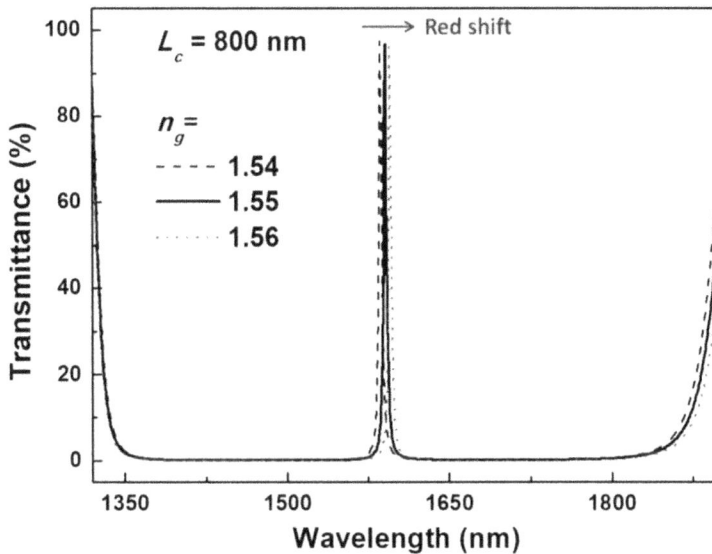

Figure 5.11. Transmittance spectra of a waveguide-based cavity for different modal index values.

refractive index of the glass fiber is about 0.01. In other words, the peak wavelength of the longitudinal mode of a DBR-based glass fiber cavity can be increased from 1590.5 to 1595 nm when the temperature is increased from 300 K to 1300 K. Therefore, a DBR-based glass fiber cavity can be used as a temperature sensor at high temperatures. The temperature sensitivity of a DBR-based glass fiber cavity can be defined as equation (5.5b). The calculated temperature sensitivity is 4.5 pm K^{-1}.

$$n(T) = n_0 + T(\mathrm{d}n/\mathrm{d}T) \tag{5.5a}$$

$$\text{Sensitivity} = \Delta\lambda_l/\Delta T \tag{5.5b}$$

5.3 Nanoplasmonic sensors

Surface plasmon polariton (SPP) waves propagate at metal/dielectric interfaces owing to the generation of collective free electrons at metal surfaces. The propagation characteristics of SPP waves are sensitive to the dielectric constants of the metal and the dielectric medium. In a plasmonic waveguide, the modal index is higher than the refractive index of the dielectric medium due to the localized EM waves. Figure 5.12 plots the cross-sectional structures of various long-range plasmonic waveguides. The modal index and propagation loss of these plasmonic waveguides can be calculated using the effective index method [5]. In coupled rib, V-groove, and wedge plasmonic waveguides, smaller values of θ result in higher modal indices and larger propagation losses owing to the stronger field localization effect. In coupled rib and metal–dielectric–metal plasmonic waveguides, smaller values of W result in higher modal indices and larger propagation losses, which are due to more strongly coupled SPP waves [6]. The widely used metals are Au, Ag, Al, and

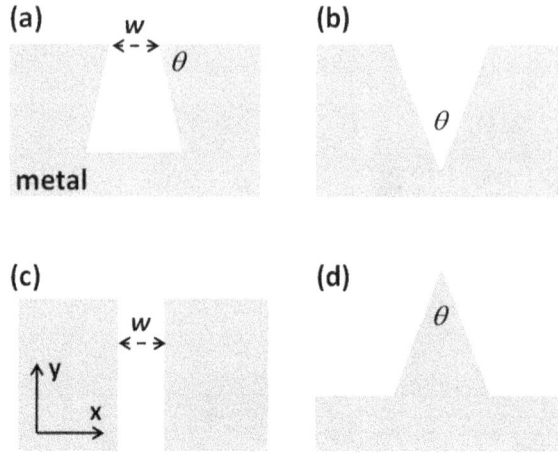

Figure 5.12. Cross-sectional structures of various plasmonic structures. (a) Coupled ribs; (b) V-groove; (c) metal–dielectric–metal; and (d) wedge.

Cu owing to their relatively lower ohmic loss. Because of its chemical stability, Au is used to fabricate plasmonic waveguides in order to realize plasmonic components such as Y-splitters, Y-combiners, ring resonators, and Mach–Zehnder interferometers [7]. Ring resonators and Mach–Zehnder interferometers are both related to wave interference, which can form on-state and off-state signals via constructive and destructive interference, respectively. In ring resonators, a larger Q value corresponds to a narrower peak width in the transmission spectrum and thus an increase in the number of wavelength channels within the FSR. Therefore, propagation losses in plasmonic waveguides must be minimized in order to increase the Q values of resonators and interference devices.

Conceptually, the Q value of a cavity or resonator is related to the stored energy ($E = qh\upsilon$) and the energy loss in one cycle ($E_l = (-dE/dt)(1/\upsilon)$), which can be written as equation (5.6a), where q is the number of photons and υ is the photon frequency. The rate equation for the number of photons (q) can be written as equation (5.5b), where R is a constant. After the rate equation has been solved, the function $q(t)$ can be written as equation (5.5c), where q_0 is the initial number of photons and τ_c is a characteristic time which equals the reciprocal of R. By inserting equation (5.5c) into equation (5.5a), the quality factor can be expressed as equation (5.5d), where $\Delta\nu_c$ is the bandwidth of the resonant mode. Therefore, the Q value of a resonator can be determined by analyzing its transmittance spectrum. The Q value is 3×10^7 when the resonant wavelength (λ_0) and characteristic time of the cavity are 1550 nm and 155 ns, respectively. In figure 5.11, the Q (τ_c) value is about 1600 (1.36×10^{-12} s), which can be computed by using equation (5.5d). The short τ_c is due to the short optical path length of the waveguide-based cavity.

$$Q = 2\pi\frac{E}{E_l} = 2\pi\frac{q\upsilon}{-dq/dt} \tag{5.5a}$$

$$\mathrm{d}q/\mathrm{d}t = -Rq \tag{5.5b}$$

$$q = q_0 \exp(-t/\tau_c) \tag{5.5c}$$

$$Q = 2\pi\nu\tau_c = \nu/\Delta\nu_c \tag{5.5d}$$

When considering the losses caused by internal and external mechanisms, the quality factor can be computed using equation (5.6a), where q_i and q_{ext} are the numbers of photons which participate in the internal and external loss processes, respectively. In other words, there are two characteristic times which can be used to compute the internal quality factor (Q_i) and the external quality factor (Q_{ext}), respectively. Equation (5.6a) can then be written in a simple form, as shown in equation (5.6b). When the propagation loss can be ignored, the coupling loss dominates the Q value of a cavity. In other words, Q can approach Q_{ext} in a low-loss cavity. On the other hand, equation (5.5d) shows that larger Q values correspond to longer τ_c values. In other words, there is a trade-off between the sensitivity to the refractive index and the response time (characteristic time) of a dielectric waveguide-based cavity.

$$\frac{1}{Q} = \frac{-\mathrm{d}q_i/\mathrm{d}t}{2\pi\nu q_i} + \frac{-\mathrm{d}q_{\mathrm{ext}}/\mathrm{d}t}{2\pi\nu q_{\mathrm{ext}}} \tag{5.6a}$$

$$\frac{1}{Q} = \frac{1}{Q_i} + \frac{1}{Q_{\mathrm{ext}}} \tag{5.6b}$$

The combination of a dielectric waveguide, a plasmonic waveguide, and a dielectric waveguide can form a resonator that can detect changes in the refractive index, as shown in figure 5.13. The gap width of the MDM plasmonic waveguide and the width of the dielectric slab waveguide are both W. The refractive index of the guiding (cladding) layer is 3.5 (1.0). Si can be used as the guiding layer in the long-range communication window centered at a wavelength of 1550 nm. To excite SPP waves, the TM mode of the slab waveguide must be excited. Equations (5.1a), (5.1b), (5.1c), and (5.3c) can be used to compute the propagation constant (modal index) of the dielectric slab waveguide when the W value and operational

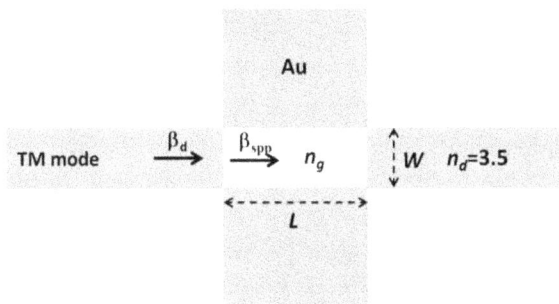

Figure 5.13. Dielectric/plasmonic/dielectric waveguide-based sensor.

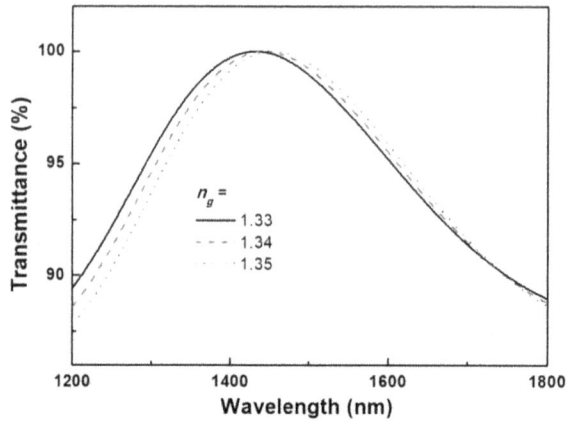

Figure 5.14. Transmittance spectra of a dielectric/plasmonic/dielectric waveguide-based sensor.

wavelength (λ_0) are known. The calculated modal index of the Si slab waveguide is 2.507 at $\lambda_0 = 1550$ nm when the W value is 300 nm. Equation (3.12) can be used to compute the propagation constant (modal index) of the MDM plasmonic waveguide when the W value, n_g value, and operational wavelength (λ_0) are known. The calculated modal index of the Au/dielectric/Au plasmonic waveguide is 1.432 at $\lambda_0 = 1550$ nm when the W value and n_g value are 300 nm and 1.33, respectively. The modal index values of the Au/dielectric/Au plasmonic waveguide are 1.432, 1.443, and 1.454 when the n_g values are 1.33, 1.34, and 1.35, respectively.

A dielectric/plasmonic/dielectric waveguide can be viewed as an $n_H/n_L/n_H$ structure, where n_H and n_L are the modal index values of the dielectric waveguide and the plasmonic waveguide, respectively. Equations (2.11) and (2.12) can be used to compute the transmittance and reflectance spectra of the effective $n_H/n_L/n_H$ structure. The transmittance spectra of the $n_H/n_L/n_H$ structure are plotted in figure 5.14 for a plasmonic waveguide length of 1000 nm. It should be noted that n_H and n_L can be assumed to be wavelength independent in the wavelength range from 1200 to 1800 nm, owing to the low dispersion characteristics of the dielectric waveguide and the plasmonic waveguide. In the transmittance spectra, the peak wavelength increases from 1432 to 1454 nm as the n_g value increases from 1.33 to 1.35. The sensitivity to changes in the refractive index (S) is given by the relationship $S = \Delta\lambda_p/\Delta n_g$, where $\Delta\lambda_p (=22$ nm) is the change in the peak wavelength and $\Delta n_g (=0.02)$ is the change in n_g. Therefore, the calculated S value is 1100. Conceptually, the S value is related to the W and L of the plasmonic waveguide. In other words, the sensitivity of a dielectric/plasmonic/dielectric waveguide-based sensor can be optimized by varying W and L.

Bibliography

[1] Pollock C R and Lipson M 2003 *Integrated Photonics* (Dordrecht: Kluwer Academic Publisher)
[2] Villar I D, Hernaez M, Zamarreno C R, Sanchez P, Fernandez-Valdivielso C, Arregui F J and Matias I R 2012 Design rules for lossy mode resonance based sensors *Appl. Opt.* **51** 4298–307

[3] Mihailov S J, Grobnic D, Hnatovsky C, Walker R B, Lu P, Coulas D and Ding H 2017 Extreme environment sensing using femtosecond laser-inscribed fiber Bragg gratings *Sensors* **17** 2909

[4] Tan C Z and Arndt J 2000 Temperature dependence of refractive index of glassy SiO2 in the infrared wavelength range *J. Phys. Chem. Solids* **61** 1315–20

[5] Bozhevolnyi S I 2006 Effective-index modeling of channel plasmon polaritons *Opt. Express* **14** 9467–76

[6] Liao Y-S, Wu J-R, Thakur D, Hsu J-S, Dwivedi R P and Chang S H 2022 Power loss reduction of angled metallic wedge plasmonic waveguides via the interplay between near-field optical coupling and modal coupling *Photonics* **9** 663

[7] Bozhevolnyi S I, Volkov V S, Devaux E, Laluet J-Y and Ebbesen T W 2006 Channel plasmon subwavelength waveguide components including interferometers and ring resonators *Nature* **440** 508–11

IOP Publishing

Light–Material Interactions and Applications in Optoelectronic Devices

Anjali Chandel and Sheng Hsiung Chang

Chapter 6

Excitons

In this chapter, the characteristics of excitons in semiconductors are mathematically and graphically described. To understand the relation between the coulomb potential energy and the exciton binding energy, the exciton radius (Bohr radius) in Si is computed using the coulomb potential energy formula or the Bohr radius formula, which shows that the discrepancy originates from the reduced mass, thereby explaining the wave behavior of energetic electrons in quantized orbits. In addition, the exciton generation, dissociation, radiative emission, diffusion, and energy transfer are described and discussed by analyzing the governing equations and/or experimental data. The theoretical prediction shows that the time-dependent carrier density is related to the initial carrier intensity via different relaxation pathways, including exciton–exciton annihilation, energy transfer, exciton radiative emission, and defect-mediated relaxation. In other words, the exciton dynamics can be explored by analyzing the time-resolved photoluminescence (PL) curves.

6.1 Excitons in semiconductors

Excitons are charged quasiparticles. One exciton consists of one electron in the conduction band (the lowest unoccupied molecular orbital) and one hole in the valence band (the highest occupied molecular orbital) of inorganic (organic) semi-conductors. In real space, there is a distance between the electron and the hole, which is defined as the exciton radius. According to Coulomb's law, the electrical potential energy between the electron and hole can be computed when the dielectric constant and exciton radius are known. The electrical potential energy can be viewed as the exciton binding energy when the shapes of the electron and hole clouds are considered in the calculation [1].

Let us illustrate the properties of excitons by discussing the electrical potential energy, which is computed using Coulomb's law. The electrical potential energy (U)

between one electron and one hole can be computed using equation (6.1), where e is the electrical charge of an electron, ε_0 is absolute permittivity, ε_d is relative permittivity, and r is the distance between the electron and the hole. In high-refractive-index materials, the high ε_d is due to the dielectric screening effect. In addition, the dielectric screening effect also results in large values of r. Therefore, the U values of high-refractive-index materials are low, mainly owing to the high dielectric screening effect. In other words, the electron and hole are easily delocalized in high-refractive-index materials, leading to weak exciton binding energy (E_b). When the exciton binding energy is lower than the thermal energy ($K_B T$), excitons can self-dissociate to form free electrons and free holes in high-refractive-index materials. K_B and T are the Boltzmann constant and the temperature, respectively. At room temperature, the thermal energy is about 25 meV. Let us try to compute the exciton radius (r) in a Si crystal using equation (6.1). The calculated r value is about 12.29 nm when the dielectric constant and the exciton binding energy of Si are 11.7 and 0.01 eV, respectively. However, the calculated r value is larger than the exciton Bohr radius of a Si crystal, which is about 5 nm [2]. The discrepancy between the calculated value of r and the exciton Bohr radius shows that the shapes of the electron and hole clouds greatly influence the effective distance between the centers of the electron and hole clouds. In other words, the above-mentioned classical description does not include the wave properties of charged particles in semiconductors.

$$U = \frac{e^2}{4\pi\varepsilon_0\varepsilon_d r} \tag{6.1}$$

The Bohr radius of an atom can be computed using equation (6.2a), where $\hbar$ is the reduced Planck constant, m_e is the electron mass, and m_r is the reduced mass. Here, $4\pi\varepsilon_0\hbar^2/(m_e e^2)$ is the Bohr radius of a hydrogen atom, which equals 5.29×10^{-11} m. Therefore, the exciton Bohr radius of semiconductor materials is related to the relative permittivity (ε_d) and the ratio of m_e to m_r. The ε_d and m_e/m_r values of crystalline Si materials are 11.7 and 10, respectively. Therefore, the calculated exciton Bohr radius of Si is about 6.19 nm, which is close to the reported values [2]. This shows that equation (6.2a) can be used to correctly compute the exciton radius of a semiconductor when the correct values of ε_d and m_e/m_r are known.

$$r_B = \frac{4\pi\varepsilon_0\hbar^2}{m_e e^2}\left(\frac{\varepsilon_d m_e}{m_r}\right) \tag{6.2a}$$

$$m_r = \frac{m_e^* m_h^*}{m_e^* + m_h^*} \tag{6.2b}$$

We now discuss the discrepancy between r and r_B further. In equation (6.1), the excited electron and hole are assumed to be located in a uniform dielectric medium (ε_d). Therefore, the binding energy (U) between the electron and hole is inversely proportional to r and ε_d. However, excited electrons and holes in semiconductors are not two static charged particles. In the Bohr model, electrons can stay in discrete orbitals centered on a positively charged nucleus, which indicates that the energetic electrons are viewed as energy in the form of waves. In other words, the calculated Bohr radius of a hydrogen atom is the result of the requirement for the energetic electron to simultaneously satisfy Coulomb's law between the two charged particles and its standing-wave properties in the quantized orbit. In semiconductors, the formation of chemical bonds between atoms shows that the electrons in the outer orbits can be shared with adjacent atoms due to orbital hybridization. In other words, the excited electrons (holes) in the hybrid orbitals have different effective masses. The effective masses of excited electrons and holes can be used to compute the reduced mass using equation (6.2b). Therefore, the exciton Bohr radius in semiconductors is dominated by the dielectric screening effect and the potential energy distribution. The dielectric screening effect and the potential energy distribution in semiconductors determine ε_d and m_r, respectively.

6.2 Exciton generation

Semiconductors can be classified into p-type, intrinsic-type, and n-type materials, which can absorb EM waves when the photon energy is higher than the absorption bandgap (E_g), thus forming excitons. In other words, the electrons are optically excited from E_{VBM} to E_{CBM}, thereby leaving holes in E_{VBM} when the photon energy equals E_g, as shown in figure 6.1(a) (here, E_{VBM} and E_{CBM} denote the energy levels of the valence band maximum and the conduction band minimum, respectively). In n-type (p-type) semiconductors, the energy of the photoexcited electrons (holes) can be transferred to adjacent electrons (holes) in the conduction band (valence band), thereby increasing the exciton radius due to the carrier delocalization effect. Conceptually, a larger exciton radius results in lower exciton generation efficiency (light absorption coefficient). In three-dimensional (3D) semiconductors, the absorption coefficient spectrum can be computed using equation (6.3a) [3]; it is related to the exciton Bohr radius (r_B) and the Rydberg energy (E_{Ry}). The photon energy ($\hbar\omega$) must be larger than the absorption bandgap (E_g) because the absorption coefficient is a real number. The value of E_{Ry} can be computed using equation (6.3b). The calculated E_{Ry} of crystalline Si materials is about 1 eV when ε_d and m_r are 11.7 and 0.1 m_e, respectively. The form of the function S_{3D} is written as equation

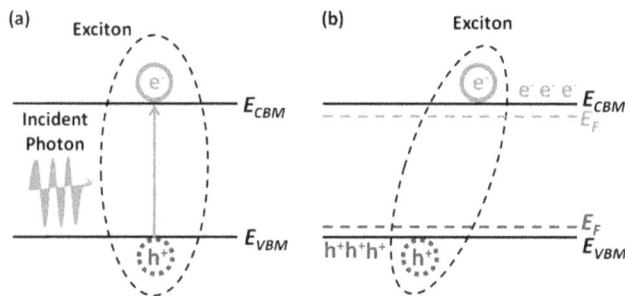

Figure 6.1. (a) Energy diagram of a photon-excited exciton in a semiconductor. (b) Energy diagram of an exciton in an n-type or p-type semiconductor.

Table 6.1. Absorption coefficient (α), reduced mass (m_r), and relative dielectric constant (ε_d) of GaN, GaAs, and GaSb at $T = 300$ K.

Compounds	α ($10^4 \times$ cm^{-1})/λ (nm)	m_r (m_e)	ε_d	References
GaN	~9.5/~363	0.1300	8.9	[4–6]
GaAs	~0.54/~875	0.0595	12.9	[7, 8]
GaSb	~0.39/~1698	0.0364	15.7	[7, 9]

(6.3c). In equation (6.3a), the first and second terms in the bracket represent the exciton transition and the continuum band transition, respectively; these determine the shape of the absorption curve. The coefficient $A_0/(2\pi^2 E_{Ry} r_B^3)$ determines the absorption strength, where A_0 is a constant which can be determined by curve fitting. In other words, the absorption coefficient is proportional to m_r^2/ε_d. To confirm this relationship, the absorption coefficients at the exciton transition peak, m_r values, and ε_d values of Ga-based III–V direct-bandgap semiconductors are listed in table 6.1. Larger m_r^2/ε_d values correspond to higher absorption coefficients at the exciton transition peak. It should be noted that GaP is not discussed owing to its indirect bandgap transition.

$$\alpha = \frac{A_0}{2\pi^2 E_{Ry} r_B^3}\left[4\pi\sum_{n=1}^{\infty}\frac{1}{n^3}\delta\left(\frac{\hbar\omega - E_g}{E_{Ry}} + \frac{1}{n^2}\right) + S_{3D}\left(\frac{\hbar\omega - E_g}{E_{Ry}}\right)\sqrt{\frac{\hbar\omega - E_g}{E_{Ry}}}\right] \quad (6.3a)$$

$$E_{Ry} = \frac{m_r e^4}{2\hbar^2(4\pi\varepsilon_0\varepsilon_d)^2} \quad (6.3b)$$

$$S_{3D}(x) = \frac{2\pi/\sqrt{x}}{1 - e^{-2\pi/\sqrt{x}}} \quad (6.3c)$$

We now discuss the effect of carrier concentration on the absorption strength of semiconductors. In figure 6.1(b), the additional electrons (holes) in the conduction band (valence band) result in delocalized excitons due to energy transfer from the excitons to the adjacent carriers, which increases the exciton radius. According to equation (6.2a), a larger exciton radius (r_B) corresponds to a smaller reduced mass (m_r), which thus results in lower absorption strength in semiconductors. In other words, a higher carrier concentration results in lower absorption strength in semiconductors. The carrier-concentration-dependent light absorption strength is the main reason that the active layers of solar cells and photodetectors are intrinsic semiconductors.

6.3 Exciton dissociation and emission

In the energy diagram of semiconductors, the electron and hole of an exciton stay in the conduction band and the valence band, respectively, and retain an attractive energy. We now discuss exciton dissociation and emission in the widely investigated organic–inorganic semiconductors, such as $CH_3NH_3PbI_3$ (methylammonium lead iodide, $MAPbI_3$), $MAPbBr_3$, $HC(NH_2)_2PbI_3$ (formamidinium lead iodide, $FAPbI_3$), and $FAPbBr_3$. When light is absorbed, electrons are excited from the valence band to the conduction band, which results in the creation of electron–hole pairs (excitons). In lead triiodide perovskites, the photoexcited electrons and holes mainly occupy lead (Pb) and iodide (I) atoms, respectively [10]. In other words, the energies of the electrons and holes can flow through the Pb and I atoms, respectively. It should be noted that the MA (FA) cations play an important role as dipoles that spatially separate the electron and hole flows, which has been theoretically predicted and experimentally demonstrated [11, 12]. In other words, the MA (FA) cations do not participate in the excitonic transition, which is due to the van der Waals interaction between MA cations and PbI_6 octahedra. The weak van der Waals interaction between the MA cations and PbI_6 octahedra results in optically excited aligned organic dipoles, which can be used to explain the high fill factor (FF) of the resultant perovskite solar cells. The FF values of lead triiodide perovskite solar cells can be higher than 80% under one sun of illumination (100 mW cm^{-2}). It should be noted that the active layer of perovskite solar cells is a polycrystalline material. In other words, defects do not effectively trap the photogenerated electrons and/or holes in polycrystalline perovskite thin films, leading to efficient exciton dissociation in lead triiodide perovskites, such as $MAPbI_3$ and $FAPbI_3$. In lead tribromide perovskites ($MAPbBr_3$ and $FAPbBr_3$), the van der Waals interaction between the MA (FA) cations and $PbBr_6$ octahedra is stronger. Therefore, optically excited aligned organic cations cannot effectively form in lead tribromide perovskites, which reduces exciton dissociation and thereby results in stronger emission due to the carrier localization effect. In other words, higher (lower) exciton dissociation results in weaker (stronger) light emission in three-dimensional crystals.

To quantitatively evaluate the exciton dissociation and emission in lead trihalide perovskites, the fraction of charge carriers divided by the excitation density (x) can be computed using equation (6.4a) [13], where n is the excitation density, m_r is the reduced mass, K_BT is the thermal energy, $\hbar$ is the reduced Planck constant, and E_b is the exciton binding energy. Equation (6.4a) can be formulated as a quadratic equation when the values of n, m_r, k_BT, $\hbar$, and E_b are known, as shown in equation (6.4b). The solution for x is $x = [-A + (A^2 + 4A)^{1/2}]/2$ because x must be a positive value. The value of A equals 14 806 when the n, m_r, k_BT, and E_b values of a lead trihalide perovskite are 1×10^{20} m^{-3}, 0.15 m_e, and 25 and 50 meV, respectively. The x value then equals one. When the value of n increases from 1×10^{20} m^{-3} to 1×10^{25} m^{-3}, the value of x decreases from one to 0.636. In other words, the efficiency with which photon energy is converted to electricity cannot reach 100% at high excitation densities, which limits the applications of solar cells and photodetectors.

$$\frac{x^2}{1-x} = \frac{1}{n}\left(\frac{m_r K_B T}{\hbar^2}\right)^{3/2} \exp\left(\frac{-E_b}{K_B T}\right) = A \qquad (6.4a)$$

$$x^2 + Ax - A = 0 \qquad (6.4b)$$

$$n = I\tau/(LE_P) \tag{6.4c}$$

Figure 6.2 plots the x–n curve when the m_r, K_BT, and E_B values of a lead trihalide perovskite are 0.15 m_e and 25 and 50 meV, respectively. The value of n can be computed using equation (6.4c), where I is the light intensity, τ is the relaxation time of the charge carriers, L is the absorption length, and E_p is the photon energy. The intensity and average photon energy of one sun with a spectrum of AM 1.5G are 100 mW cm^{-2} and 2.38 eV, respectively. The charge carrier relaxation time and absorption length of lead trihalide perovskites are about 50 ns and 500 nm, respectively. Thus, the calculated n value is 6.245×10^{20} m^{-3}, which shows that the absorbed light almost forms charge carriers in lead trihalide perovskites under the illumination of one sun (100 mW cm^{-2}). It should be noted that the x value is still larger than 95% when lead trihalide perovskites are excited under illumination of 100 suns, which shows that lead trihalide perovskites can be used as an efficient light-absorbing material in concentrated solar cells.

In two-dimensional (2D) materials, the exciton dissociation and emission can be evaluated using equation (6.5a) [14], where n_{eh} is the areal density of the charge carriers, n_{exc} is the areal density of the excitons, m_r is the reduced mass, K_BT is the thermal energy, E_b is the exciton binding energy, and $\hbar$ is the reduced Planck constant. After light absorption, the areal density of absorbed photons (N) equals the sum of n_{eh} and n_{exc}, which can be formulated as equation (6.5b). The calculated value of n_{eh} is far larger than the given n_{exc} when the E_b value is not much larger than the thermal energy at room temperature. Therefore, it almost completely forms charge carriers in 2D materials upon light absorption at low excitation densities. By analyzing equations (6.5a) and (6.5b), the influences of E_b and temperature (T) on n_{eh} can be obtained. When T (E_b) approaches zero (infinity), n_{eh} approaches zero. In

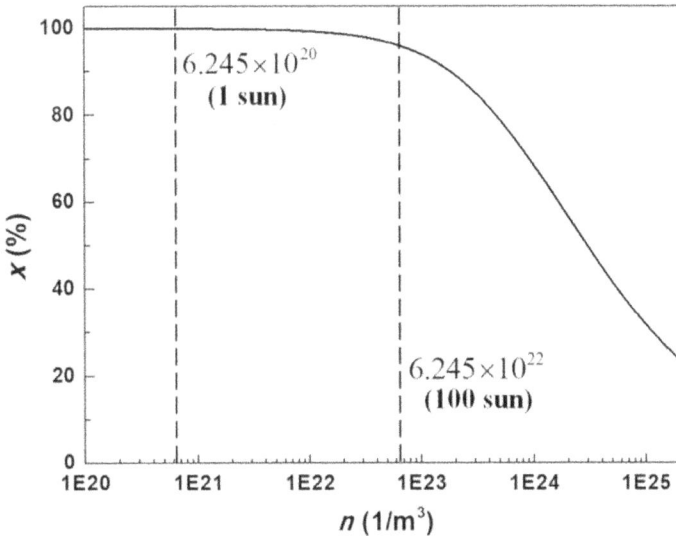

Figure 6.2. The n-dependent fraction of charge carriers versus excitation density.

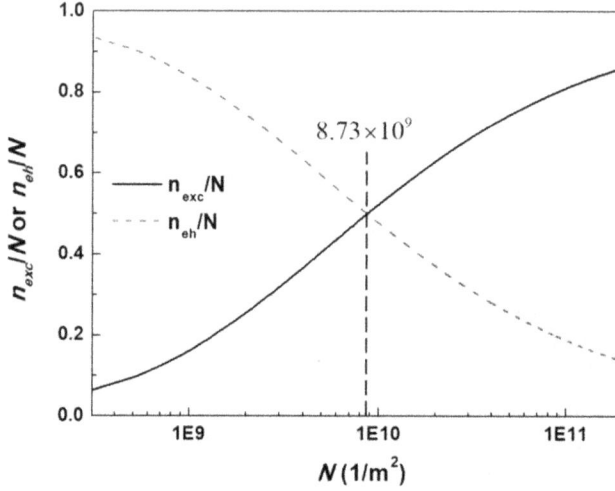

Figure 6.3. N-dependent n_{exc}/N and n_{eh}/N ratios for m_r, $k_B T$, and E_B values of $0.1m_e$, 25 meV, and 350 meV, respectively.

other words, lower values of T and larger values of E_B result in smaller values of n_{eh} and therefore larger values of n_{exc}, which shows that an efficient light-emitting diode must have a large value of E_b in the emissive layer when operating at low temperatures.

$$n_{eh} = \sqrt{n_{exc} \frac{m_r K_B T}{2\pi \hbar^2} \exp\left(\frac{-E_b}{K_B T}\right)} \quad (6.5a)$$

$$N = e_{eh} + e_{exc} \quad (6.5b)$$

Figure 6.3 plots the n_{exc}/N–N and n_{eh}/N–N curves for m_r, $K_B T$, and E_b values of emissive 2D materials of $0.1\,m_e$ and 25 and 350 meV, respectively. When N is 8.73×10^9 m^{-2}, the absorbed photons form 50% excitons and 50% charge carriers in emissive 2D materials. The n_{exc}/N ratio can be increased to 100% by increasing N, which can be explained by the increase in free electron–hole capture events that form radiative excitons. Therefore, emissive 2D materials such as WS$_2$ [15] and quasi-2D perovskites can be used as efficient gain media in laser systems [16].

6.4 Exciton diffusion and energy transfer

In organic photovoltaics (OPVs), exciton diffusion and energy transfer greatly influence the formation of photocurrents owing to the large exciton binding energy (E_b) of organic materials. In p-type organic materials, the diffusion length of excitons is shorter than the absorption length of light waves. The diffusion length of excitons can be computed using equation (6.6a), where D is the diffusion coefficient and τ is the intrinsic exciton lifetime. D can be computed using equation (6.6b), where K is the bimolecular annihilation rate constant and R_a is the annihilation radius of singlet excitons. The τ and K values can be obtained by fitting the time-resolved PL curve or transient absorbance curve to the rate equation of the charge carriers, which can be written as equation (6.6c). The rate equation is a first-order differential equation. The solution of the differential equation can be written as equation (6.6d), where N_0 is the initial density of charge carriers after the light excitation.

$$L_D = \sqrt{D\tau} \tag{6.6a}$$

$$D = K/(4\pi R_a) \tag{6.6b}$$

$$dn(t)/dt = -(1/\tau)n(t) - Kn^2(t) \tag{6.6c}$$

$$n(t) = N_0 e^{-t/\tau}/[1 + K\tau N_0(1 - e^{-t/\tau})] \tag{6.6d}$$

In p-type organic materials, the R_a value is about 1 nm. When the exciton lifetime (τ) and diffusion length (L_D) are 500 ps and 5 nm, respectively, the diffusion coefficient (D) can be computed using equation (6.6a). This allows the bi-molecular annihilation rate constant (K) to be computed using equation (6.6b). Bimolecular annihilation is also called exciton–exciton annihilation, which describes non-radiative recombination. The calculated D and K values are 5×10^{-4} cm^2 s^{-1} and $20\pi \times 10^{-11}$ cm^3 s^{-1}, respectively, which are used as the fixed parameters of the $n(t)$ curves in figure 6.4. When the N_0 value is 1×10^{18} cm^{-3}, the log(n)–t curve is almost a straight line, which shows that carrier relaxation undergoes exponential decay. In other words, bimolecular annihilation can be ignored when the excitation intensity is low. The calculated $K\tau N_0$ value is 0.1π, which is less than one, thereby explaining the straight line. When the N_0 value is 1×10^{20} cm^{-3}, the log(n)–t curve shows a

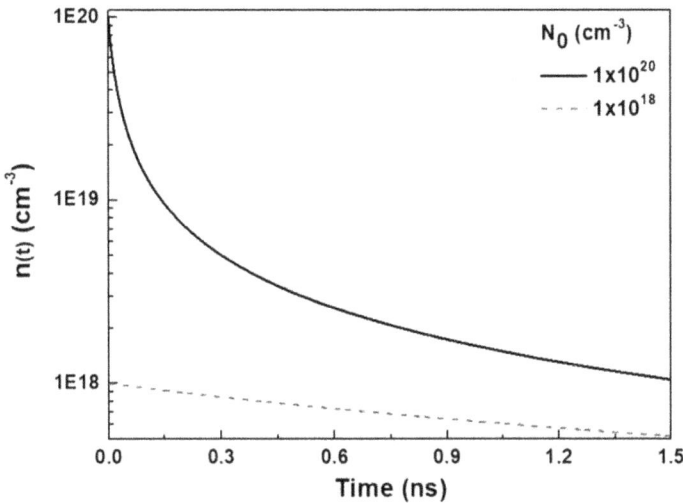

Figure 6.4. $n(t)$ curves for τ and K values of 500 ps and $20\pi \times 10^{-11}$ cm^3 s^{-1}, respectively.

distinguishable graded slope before 1.2 ns, which indicates that bimolecular annihilation significantly influences the effective relaxation time when the carrier density (n) is larger than 2×10^{18} cm^{-3}. Let us estimate the excitation density of one sun for p-type organic materials using equation (6.4c). The calculated excitation density is about 6.6×10^{12} cm^{-3} when the absorption length (L), relaxation time (τ), and photon energy (E_p) are 200 nm, 0.5 ns, and 2.38 eV, respectively. In other words, sunlight-generated carriers in p-type organic materials would not undergo an observable bimolecular annihilation process under illumination of one sun, which partially explains the efficient photocurrent generation that takes place in organic photovoltaic and photodetector devices.

Let us try to predict the time-resolved PL ($n(t)$) curve of a photoexcited MAPbI$_3$ material using equations (6.4a) and (6.6a). The parameters used are listed in table 6.2. The given light intensity (I), exciton lifetime (τ), absorption length (L), and photon energy (E_p) values can be used to compute N_0, which is 6.25×10^{14} cm^{-3}. The given τ, R_a, and L_D values can be used to compute $n(t)$. It should be noted that the exciton Bohr radius is assumed to be R_a. The calculated $n(t)$ curve is plotted in figure 6.5. The charge carriers (n) decay rapidly to 1% within 12.1 ns owing to the

Table 6.2. Light intensity (I), exciton lifetime (τ), absorption length (L), photon energy (E_p), diffusion coefficient (D), and annihilation radius of singlet excitons (R_a).

Parameter	I (mW cm^{-2})	τ (ns)	L (nm)	E_p (eV)	R_a (nm)	D (cm^2 s^{-1})
Value	100	50	500	2.38	4.57 [17]	2 [18]

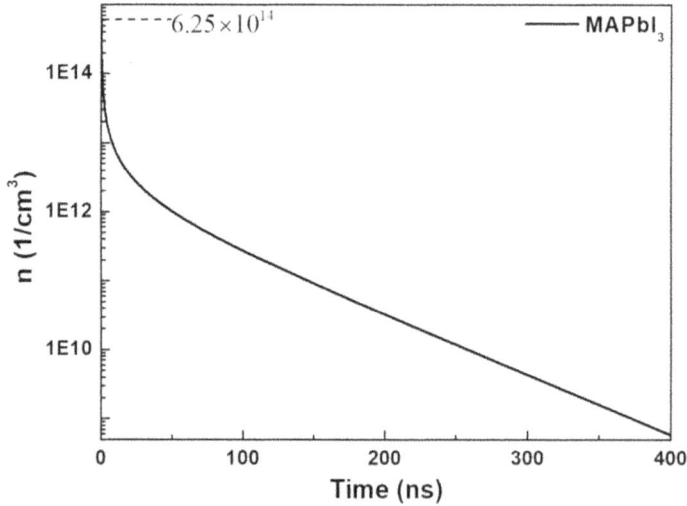

Figure 6.5. $n(t)$ curves of a photon-excited MAPbI$_3$ film for N_0, τ, and K values of 6.25×10^{14} cm^{-3}, 50 ns, and 5.74×10^{-6} cm^3 s^{-1}, respectively.

strong bimolecular annihilation process. After 50 ns, the $\log(n)$–t curve is almost a straight line with a fixed exponential decay constant. It seems that the sunlight-generated carriers decay almost non-radiatively in the $MAPbI_3$ film, which does not explain the efficiency of $MAPbI_3$ perovskite solar cells. In a $MAPbI_3$ perovskite film, the photogenerated electrons and holes are spatially separated by MA^+ cations [11, 12], which suppresses the formation of excitons or electron–hole pairs and thereby results in relatively lower exciton–exciton annihilation. The dissociated excitons in $MAPbI_3$ films efficiently form free electrons and free holes, which explains the high photocurrent density values of the resultant solar cells.

In $MAPbI_3$ solar cells, the electron transport layer (ETL) and hole transport layer (HTL) can effectively collect photogenerated electrons and photogenerated holes, respectively, which results in short electron–hole pair (exciton) lifetimes in $MAPbI_3$ perovskites. The lifetime of the residual excitons or electron–hole pairs in the active layer can be shorter than 10 ns, as shown in figure 6.6. In efficient $MAPbI_3$ solar cells, phenyl-C61-butyric acid methyl ester doped with bathocuproine (BCP:PCBM) and P3CT-Na modified indium tin oxide (ITO) are used as the ETL and the HTL, respectively. The normalized PL intensity rapidly decreases, reaching 1.7% within 1.625 ns due to exciton–exciton annihilation and/or short carrier collection times at the $MAPbI_3$/ETL and HTL/$MAPbI_3$ interfaces under pulsed laser excitation. In general, the carrier collection time from the light-absorbing layer to the ETL or HTL in an efficient solar cell can be shorter than 10 ps owing to the strong driving force at the interfaces [19]. It should be noted that the prominent peak at 1.175 ns is a feature of the pulsed laser, which indicates that the relaxation time in the exciton–exciton annihilation process or the carrier collection process is shorter than the pulse duration of the excitation laser. After 1.625 ns, the curve is mainly related to residual

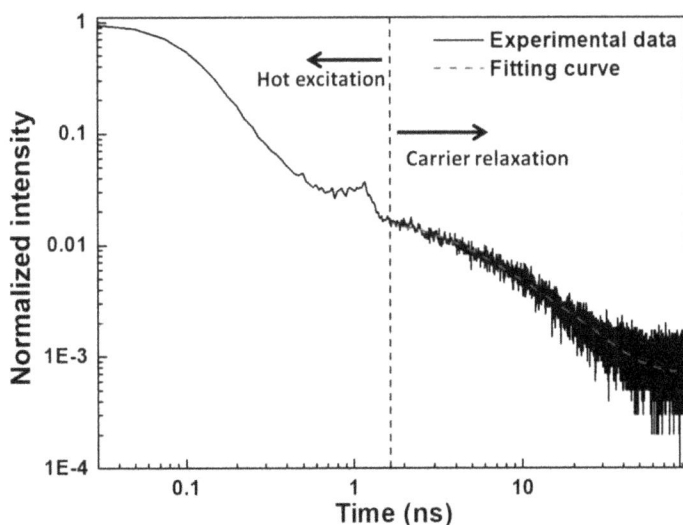

Figure 6.6. Normalized time-resolved PL curve of a BCP:PCBM/$MAPbI_3$/P3CT-Na/ITO/glass sample under pulsed laser excitation.

excitons (electron–hole pairs) in the MAPbI$_3$ film. The characteristic times of the charge carriers can be obtained by fitting the curve to equation (6.7), where I_b is the background value, A_1 is the amount of non-radiative residual excitons, τ_1 is the lifetime of the non-radiative residual excitons, A_2 is the amount of radiative residual excitons, and τ_2 is the lifetime of the radiative residual excitons. The fitting parameters are listed in table 6.3. The I_b value must be determined carefully in order to correctly compute the τ_2 value. In addition, the position of the dashed line in figure 6.7 is related to the instrument response function (IRF) of the detection system and the pulse shape of the excitation laser, which can significantly influence the calculated τ_1 value. The total of A_1 and A_2 is 1.76%, which is related to the residual excitons in the MAPbI$_3$ perovskite. In other words, most of the excitons are dissociated at the MAPbI$_3$/HTL and MAPbI$_3$/ETL interfaces, which explains the high photocurrent density values of efficient MAPbI$_3$ solar cells.

$$I(t) = I_b + A_1 e^{-t/\tau_1} + A_2 e^{-t/\tau_2} \qquad (6.7)$$

In efficient solar cells, the energy transfer times at the interfaces for electrons and holes can be shorter than 1 ps, according to analysis performed using ultrafast pump–probe techniques [20, 21]. In a perovskite/HTL sample or a perovskite/ETL sample, the quenching-mediated diffusion equation can be written as equation (6.8a) [22], where D is the diffusion coefficient and τ_q is the average carrier lifetime. In a

Table 6.3. Fitting parameters used in equation (6.7).

Parameter	I_B	A_1	τ_1 (ns)	A_2	τ_2 (ns)
Value	7.1×10^{-4}	1.18×10^{-2}	3.57	0.58×10^{-2}	14.42

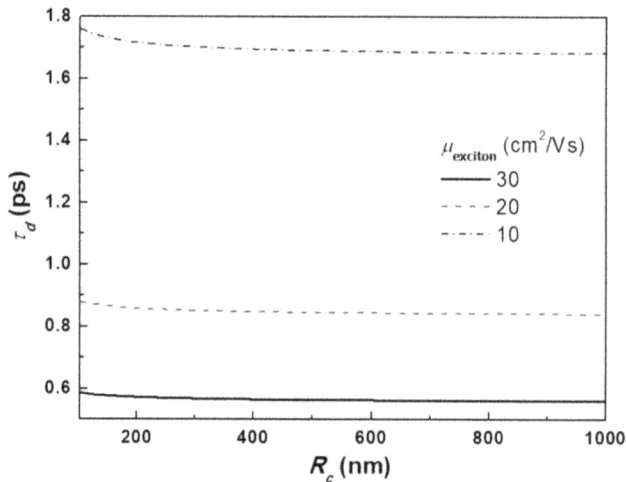

Figure 6.7. τ_d–R_C relation of MAPbI$_3$/ETL samples with different charge carrier (exciton) mobility values.

perovskite/HTL sample or a perovskite/ETL sample, the carriers mainly diffuse from the bottom region of the perovskite film to the interface, thus producing one-dimensional (1D) diffusion. The 1D diffusion length (L_D) can be computed using the simple relation $L_D = \sqrt{D\tau}$, where τ is the relaxation time of the charge carriers in the perovskite without the quenching effect. The charge carrier density can be written as equation (6.8b), where α is the absorption coefficient. By combining equations (6.8a) and (6.8b), the relation between τ_q and τ can be written as equation (6.8c).

$$\frac{\partial n(z,\ t)}{\partial t} = D\frac{\partial^2 n(z,\ t)}{\partial z^2} - \frac{n(z,\ t)}{\tau_q} \tag{6.8a}$$

$$n(z,\ t) = N_0 e^{-\alpha z} e^{-t/\tau} \tag{6.8b}$$

$$\tau_q = \tau/(1 + L_D^2 \alpha^2) \tag{6.8c}$$

We now discuss the influences of L_D and α on τ_q using equation (6.8c). In a MAPbI$_3$/HTL sample, the carrier diffusion length (L_D) has a fixed value. When the absorption coefficient (α) is larger, the photogenerated carriers in the MAPbI$_3$ film are closer to the MAPbI$_3$/HTL interface, resulting in better hole collection efficiency. Therefore, a larger absorption coefficient (α) corresponds to a shorter average carrier lifetime (τ_q). In MAPbI$_3$ films, higher crystallinity corresponds to longer L_D and larger α, thus resulting in shorter τ_q. In other words, the τ_q value of a MAPbI$_3$/HTL sample can be used to evaluate the crystallinity of the MAPbI$_3$ film.

In a MAPbI$_3$/ETL sample, the exciton (electron–hole pair) dissociation time (τ_d) at the MAPbI$_3$/ETL interface can be computed using equation (6.9a) [20], where γ equals $e\mu_{\text{exciton}}/\varepsilon_d \varepsilon_0$, r_{exciton} is the exciton radius, E_b is the exciton binding energy, J_1 is a first-order Bessel function, and F is a function of the built-in electric field (E) at the MAPbI$_3$/ETL interface. The relative permittivity (ε_d) and thermal energy ($K_B T$) also influence the F value. The E value can be computed using equation (6.9c), where E_{PL} is the energy level of photogenerated electrons in the MAPbI$_3$ film before radiation, $E_{\text{LUMO}}^{\text{ETL}}$ is the lowest unoccupied molecular orbital energy level of the ETL, and R_c is the charge transfer radius of photogenerated electrons in the MAPbI$_3$ film. In the R_c region, excitons (electron–hole pairs) may dissociate. Therefore, larger R_c values correspond to faster exciton (electron–hole pair) dissociation. Figure 6.7 plots the τ_d–R_c curves of MAPbI$_3$/ETL samples with different exciton (charge carrier) mobility values, which shows that exciton mobility plays a decisive role. The fitting parameters are listed in table 6.4.

Table 6.4. Fitting parameters used in equation (6.9a).

Parameter	ε_d [23]	r_{exciton} (nm)	E_{PL} (eV)	$E_{\text{LUMO}}^{\text{ETL}}$ (eV)	E_B (meV)	$K_B T$ (meV)
Value	34	4.57	−3.95	−4.1	20	25

The calculated results show that exciton (charge carrier) mobility greatly influences the dissociation time (τ_d). The τ_d value decreases from about 1.7 ps to about 0.58 ps as the $\mu_{exciton}$ value increases from 10 to 30 cm^2 V^{-1} s^{-1} and the R_c value ranges from 200 to 1000 nm. It should be noted that the calculated τ_d values are about 1 ps, which is consistent with experimental results [19]. On the other hand, the R_c value for the MAPbI$_3$ film of the MAPbI$_3$/ETL sample can be obtained if ε_d, $r_{exciton}$, E_{PL}, E_{LUMO}^{ETL}, E_b, $K_B T$, and τ_d are known. To effectively generate a photocurrent, the doubled length of the sum of R_c and L_D must be larger than the thickness of the light-absorbing layer of the solar cell.

$$1/\tau_d = \frac{3\gamma}{4\pi r_{exciton}^3}(e^{-E_B/K_B T})\frac{J_1(2\sqrt{-2F(E)})}{\sqrt{-2F(E)}} \tag{6.9a}$$

$$F = e^3 E/[8\pi\varepsilon_d\varepsilon_0(K_B T)^2] \tag{6.9b}$$

$$E = (E_{PL} - E_{LUMO}^{ETL})/eR_c \tag{6.9c}$$

Bibliography

[1] Nayak P K 2013 Exciton binding energy in small organic conjugated molecule *Synth. Met.* **174** 42–5

[2] von Behren J, von Buuren T, Zacharias M, Chimowitz E H and Fauchet P M 1998 Quantum confinement in nanoscale silicon: the correlation of size with bandgap and luminescence *Solid State Commun.* **105** 317–22

[3] Elliott R 1957 Intensity of optical absorption by excitons *Phys. Rev.* **108** 1384–9

[4] Muth J F, Brown J D, Johnson M A, Yu Z, Kolbas R M, Cook J W and Schetzina J F 1999 Absorption coefficient and refractive index of GaN, AlN and AlGaN alloys *MRS Internet J. Nitride Semicond. Res.* **4** 502–7

[5] Chen G D, Smith M, Lin J Y, Jiang H X, Wei S-H, Khan M A and Sun C J 1996 Fundamental optical transitions in GaN *Appl. Phys. Lett.* **68** 2784–6

[6] Levinshtein M M, Rumyantsev S L and Shur M S 2001 *Properties of Advanced Semiconductor Materials: GaN, AlN, InN, BN, SiC, SiGe* (New York: Wiley)

[7] Schaefer S T, Gao S, Webster R T, Kosireddy R R and Johnson S R 2020 Absorption edge characteristics of GaAs, GaSb, InAs, and InSb *J. Appl. Phys.* **127** 165705

[8] Sanchez T G, Perez J V, Conde P G and Collantes D P 1991 Five-valley model for the study of electron transport properties at very high electric fields in GaAs *Semicond. Sci. Technol.* **6** 862–71

[9] Bennett H S and Hung H 2003 Dependence of electron density on Fermi energy in N-type gallium antimonide *J. Res. Natl. Inst. Stand. Technol.* **108** 193–7

[10] Li W, Lu J, Bai F-Q, Zhang H-X and Prezhdo O V 2017 Hole trapping by iodine interstitial defects decreases freee carrier losses in perovskite solar cells: a time-domain *ab initio* study *ACS Energy Lett.* **2** 1270–8

[11] Frost J M, Bulter K T, Brivio F, Hendon C H, van Schilfgaarde M and Walsh A 2014 Atomistic origins of high-performance in hybrid halide perovskite solar cells *Nano Lett.* **14** 2584–90

[12] Hsu H-C *et al* 2019 Photodriven dipole reordering: key to carrier separation in metalorganic halide perovskites *ACS Nano* **13** 4402–9

[13] Zheng K, Zhu Q, Abdellah M, Messing M E, Zhang W, Generalov A, Niu Y, Ribaud L, Canton S E and Pullerits T 2015 Exciton binding energy and the nature of emissive states in organometal halide perovskites *J. Phys. Chem. Lett.* **6** 2969–75

[14] Cingolani R, Calcagnile L, Coli G and Rinaldi R 1996 Radiative recombination processes in wide-band-gap II–VI quantum wells: the interplay between excitons and free carriers *J. Opt. Soc. Am.* B **13** 1268–77

[15] Barth I *et al* 2024 Lasing from a large-area 2D material enabled by a dual-resonance metasurface *ACS Nano* **18** 12897–904

[16] Zhao C and Qin C 2020 Quasi-2D lead halide perovskite gain materials toward electrical pumping laser *Nanophotonics* **10** 2167–80

[17] Baranowski M and Plochocka P 2020 Excitons in metal-halide perovskites *Adv. Energy Mater.* **10** 1903659

[18] Scajev P, Miasojedovas S and Jursenas S 2020 A carrier density dependent diffusion coefficient, recombination rate and diffusion length in MAPbI$_3$ and MAPbBr$_3$ crystals measured under one- and two-photon excitations *J. Mater. Chem.* C **8** 10290–301

[19] Dursun I, Maity P, Yin J, Turedi B, Zhumekenov A A, Lee K J, Mohammed O F and Bakr O M 2019 Why are hot holes easier to extract than hot electrons from methylammonium lead iodide perovskite? *Adv. Energy Mater.* **9** 1900084

[20] Ohkita H and Ito S 2011 Transient absorption spectroscopy of polymer-based thin-film solar cells *Polymer* **52** 4397–417

[21] Chang S H, Chiang C-H, Cheng H-M, Tai C-Y and Wu C-G 2013 Broadband charge transfer dynamics in P3HT:PCBM blended film *Opt. Lett.* **38** 5342–5

[22] Lee E M Y and Tisdale W A 2015 Determination of exciton diffusion length by transient photoluminescence quenching and its application to quantum dot films *J. Phys. Chem.* C **119** 9005–15

[23] Dong Q, Fang Y, Shao Y, Mulligan P, Qiu J, Cao L and Huang J 2015 Electron–hole diffusion lengths >175 mm in solution-grown CH3NH3PbI3 single crystals *Science* **347** 967–9

IOP Publishing

Light–Material Interactions and Applications in Optoelectronic Devices

Anjali Chandel and Sheng Hsiung Chang

Chapter 7

Energy diagrams and carrier properties

In this chapter, the energy diagrams, carrier properties, carrier drift, and carrier diffusion are mathematically and graphically described in order to explain the electrical properties of various semiconducting materials, such as GaN, SiC, Ga_2O_3, ZnO, and organic thin films. After explaining how electrons' energy and momentum (E–k) relate to the crystal's energy potential distribution, the concept of effective mass is introduced. In addition, an energy diagram is used to define direct- and indirect-bandgap semiconductors. An energy diagram is used to illustrate the depletion region and potential barriers at metal/semiconductor interfaces, which can be used to explain the formation of ohmic or Schottky contacts. To explain the types of dopants, changes in the Fermi level of ZnO caused by oxygen vacancies, n-type doping, and p-type doping are conceptually described. Moreover, the equations governing carrier drift in inorganic semiconductors and carrier diffusion in organic semiconductors are described and discussed, which can be used to explain the working mechanisms of light-emitting diodes, solar cells, and field-effect transistors.

7.1 Energy diagrams of materials

In metals, the energy of electrons flows from the high-potential electrode (the left-hand side in figure 7.1(a)) to the low-potential electrode (right-hand side) along the R axis under an external bias when the electrons are located at the tilted Fermi level. In metals, the electron density almost equals the density of the unit cell, which means that one metal atom provides one free electron, leading to a high free electron density. The short distance between free electrons results in efficient energy coupling, which explains why metals are good conductors. In semiconductors, the energy of electrons (holes) flows from the high-potential (low-potential) electrode to the low-potential (high-potential) electrode along the R axis under an external bias when electrons (holes) propagate in the conduction band (valence band), as shown in

doi:10.1088/978-0-7503-6099-9ch7
7-1

figure 7.1(b). When the electron (hole) density is higher, the E_{Fn} (E_{Fp}) value is closer to E_C (E_V) value of a semiconductor, where E_{Fn}, E_{Fp}, E_C, and E_V are the Fermi level of free electrons, the Fermi level of free holes, the energy level of the conduction band, and the energy level of the valence band, respectively. The electrical conduction mechanism of free carriers in semiconductors is similar to that of free electrons in metals, which can be described using the Drude model.

Figure 7.2 displays the energy–momentum (E–k) diagram of a metal. At points a, b, and c, the corresponding coordinates are (k_a, E_a), (k_b, E_b), and (k_c, E_c), respectively; these follow a parabolic path. The parabolic relationship can be obtained by solving the one-dimensional steady-state Schrödinger equation. The solution of the Schrödinger equation is a wave function, which can be used to describe the E–k relation of free electrons in a metal. The wave function can be written as equation (7.1a), where the coefficients A_1 and A_2 are amplitudes and k is the momentum. The E–k relation is written in equation (7.1b), where m_e^* is the effective mass of free electrons and $\hbar$ is the reduced Planck constant. At point a, the electrons are not moving; therefore, they have zero energy and no momentum. At points b and c, the electrons are moving; therefore, they simultaneously have energy

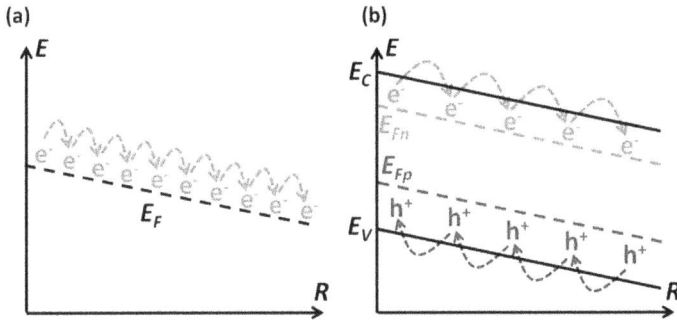

Figure 7.1. Space-dependent energy diagrams of materials subjected to an external bias. (a) Metal; (b) n-type or p-type semiconductor.

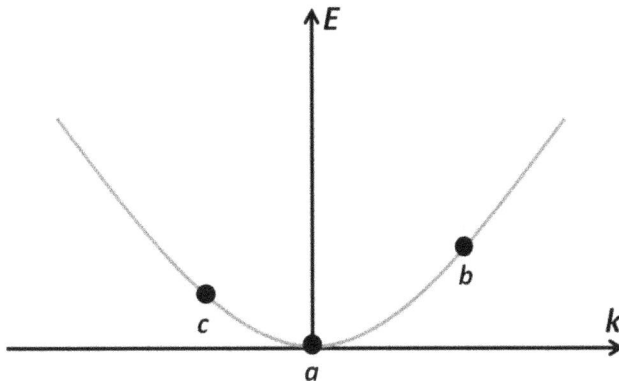

Figure 7.2. Energy–momentum (E–k) diagram of a metal.

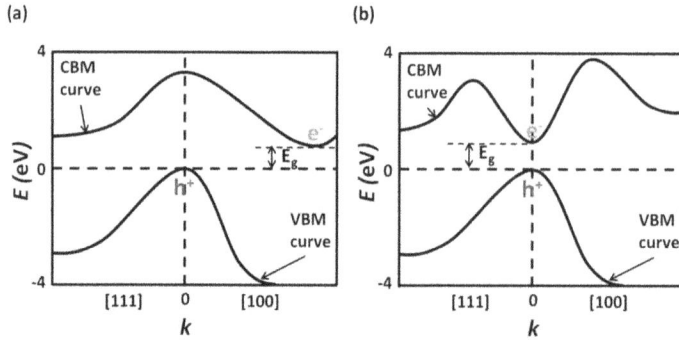

Figure 7.3. Energy–momentum (E–k) diagrams of semiconductors. (a) Indirect-bandgap; (b) direct bandgap.

and momentum according to the E–k relationship in equation (7.1b). It should be noted that the values of k can be positive or negative because k is a vector in the reciprocal space of the metal lattice.

$$\psi(x) = A_1 \cos(kx) + A_2 \sin(kx) \tag{7.1a}$$

$$k = \sqrt{\frac{2m_e^* E}{\hbar^2}} \tag{7.1b}$$

Figure 7.3 displays the E–k diagrams of an indirect-bandgap semiconductor and a direct-bandgap semiconductor. In an indirect-bandgap semiconductor, the lowest valley of the conduction band maximum (CBM) curve is not aligned with the highest peak of the valence band minimum (VBM) curve at the same k value, as shown in figure 7.3(a), which results in a non-emissive excited state. Si and GaP are two widely known indirect-bandgap semiconductors. The bandgap (E_g) is defined in the (100) crystal direction, which has the smallest difference in energy between the CBM and VBM curves. When the lowest valley of the CBM curve is aligned with the highest peak of the VBM curve at the same k value, the material is defined as a direct-bandgap semiconductor, as shown in figure 7.3(b). GaAs and GaN are two widely known direct-bandgap semiconductors. The E_g of a direct-bandgap semiconductor is defined at $k = 0$. According to equation (7.1b), a larger effective mass results in a wider parabola in the E–k diagram. Therefore, the effective mass of electrons in the lowest valley is slightly lower than that of holes at the highest peak in figure 7.3(b). In electronic devices, the injected electrons can flow into the adjacent valley when the applied voltage overcomes the potential barrier, which results in a reduction in the drift velocity and thereby decreases the operational speed.

7.2 Energy diagrams of metal/semiconductor interfaces

To effectively inject electrons (holes) from a metal into an n-type (p-type) semiconductor, an ohmic contact must be formed. The energy diagrams of a metal/n-type semiconductor interface and a metal/p-type semiconductor interface are plotted in figure 7.4, which shows ohmic contacts. The difference between the vacuum level

(a)   (b)

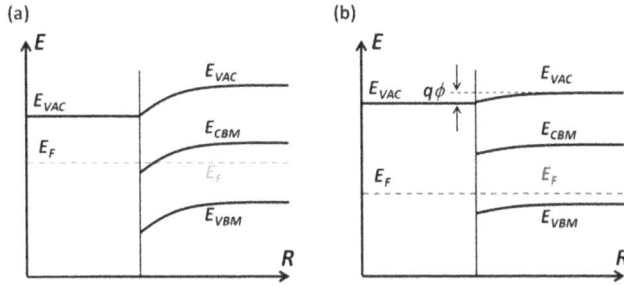

Figure 7.4. Space-dependent energy diagrams of ohmic contacts. (a) Metal/n-type semiconductor; (b) metal/p-type semiconductor.

(E_{VAC}) and the Fermi level (E_F) is defined as the work function. At the interface between a metal and an n-type (p-type) semiconductor, the Fermi level and vacuum level must be aligned. The straight dashed lines represent the Fermi levels of metals and semiconductors. To align the vacuum levels, the bands are tilted while retaining the E_g value, resulting in band bending in the junction region. The quantity of band bending equals $q\phi$, where q is the charge quantity and ϕ is the potential difference. Here, $q\phi$ is the difference between the work functions of the metal and the semiconductor. In figure 7.4(a), the Fermi level is higher than E_{CBM}, which indicates that the free electron density in the n-type semiconductor is increased due to the metal contact effect. The energy of the electrons can be injected from the metal into the n-type semiconductor with negligible contact resistance. Ag and Al are two widely used metals that can form ohmic contacts on top of n-type semiconductors. In figure 7.4(b), the difference between E_F and E_{VBM} is increased in the junction region (depletion region), which indicates that the free hole density in the p-type semiconductor is decreased due to the metal contact effect. Au and Cu are two widely used metals which can form ohmic contacts on top of p-type semiconductors. When a metal forms a contact with a semiconductor, the width of the depletion region in the semiconductor is extremely narrow, thereby allowing the energy tunneling of carriers.

At metal/p-type semiconductor interfaces, the width of the depletion region in the semiconductor can be computed using equation (7.2a), where ε_d is the relative permittivity of the semiconductor, ϕ is the quantity of potential bending, K_BT is the thermal energy, and N_a is the free hole density. When the p-type semiconductor has a large bandgap, the corresponding work function is large, resulting in a high ϕ value. Therefore, it is necessary to have a high N_a value to decrease the W_{dep} value, which can be used to form an ohmic contact at the metal/p-type semiconductor interface via the quantum tunneling effect. The contact resistance (R_C) at the metal/p-type semiconductor interface can be computed using equation (7.2b), where h is the Planck constant and m_h^* is the effective mass of the free holes in the p-type semiconductor [1]. Let us compute the W_{dep} and R_C values in p-type GaN, p-type 4H-SiC, and p-type β-Ga$_2$O$_3$ when Au is used as the metal contact. The parameters used and the calculated values are listed in table 7.1. The calculated W_{dep} values are

Table 7.1. Calculated depletion width (W_{dep}) in an Au/p-type semiconductor interface and the parameters used in the calculation. The temperature is set to 300 K. N_a is 1×10^{18} cm^{-3}.

Compound	ε_d	$q\phi$ (eV)	m_h^* (m_e)	W_{dep} (nm)	R_C (Ω cm^2)	References
c-plane GaN	9.8	1.6	0.75	41.3	1.70×10^{-8}	[2–4]
4H-SiC	10	1.9	0.66	45.6	1.70×10^{-8}	[2, 5, 6]
β-Ga$_2$O$_3$	11	2.0	0.28	49.0	1.70×10^{-8}	[2, 7]

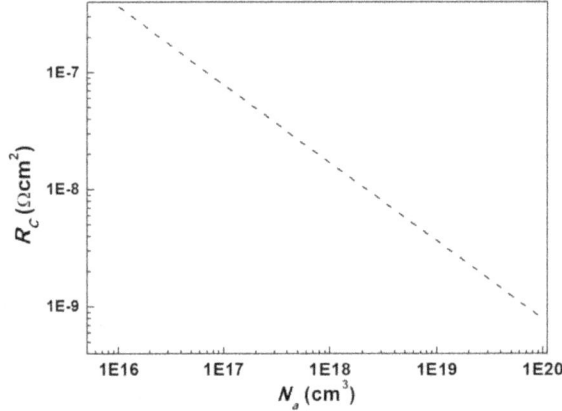

Figure 7.5. N_a-dependent contact resistance (R_C) value.

larger than 40 nm when the N_a value is fixed at 1×10^{18} cm^{-3}. The calculated R_C values are 1.70×10^{-8} Ω cm^2 because the exponent values in equation (7.2b) are close to zero, which indicates that the N_a value dominates the R_C value.

$$W_{dep} = \sqrt{|[2\varepsilon_0 \varepsilon_d (q\phi - K_B T)]/(q^2 N_a)|} \tag{7.2a}$$

$$R_C = \left(\frac{8}{9\pi N_a^2}\right)^{1/3} \frac{h}{q^2} \exp\left[\frac{4\pi}{h}(2q\phi m_h^*)^{1/2} W_{dep}\right] \tag{7.2b}$$

In other words, the p-type doping capability of large-bandgap semiconductors mainly determines the quality of the ohmic contacts made with the p-type regions of semiconductor devices, as shown in figure 7.5. When the N_a value is 1×10^{20} cm^{-3}, the R_C value can be less than 1×10^{-9} Ω cm^2. Let us evaluate the ohmic heating (H) caused by the contact resistance of a semiconductor device using the simple relation $H = I^2 R_C/A$, where I is the current injected from the metal to the p-type semi-conductor, and A is the contact area between the metal and the p-type semi-conductor. In power devices, the I value can be higher than 1000 A. The calculated H value is 1 mW when the I, R_C, and A values are 1000 A, 1×10^{-9} Ω cm^2, and 100 µm^2, respectively. The low H value indicates that the ability of semiconductors to support high dopant levels is important for semiconductor power devices.

However, higher dopant concentrations in large-bandgap semiconductors also result in lower crystallinity, thereby reducing carrier mobility. The lower carrier mobility reduces the operational speed of the resultant devices. In other words, there is a trade-off between high-power operation and high-speed operation in large-bandgap semiconductor devices.

7.3 Energy diagrams of doped semiconductors

Zinc oxide (ZnO) is a widely used semiconductor in various fields, such as optoelectronic devices, catalytic reactions, and sensors. The E_g, E_{CBM}, and E_{VBM} values of ZnO are about 3.27, -4.19, and -7.46 eV, respectively [8]. Figure 7.6 plots the energy diagrams of ZnO, ZnO with oxygen vacancies, aluminum-doped ZnO (AZO), and magnesium-doped ZnO. In ZnO, the conduction band and valence band mainly occupy the Zn and O sites, respectively [9]. In the Zn–O bonds, the electrons are transferred from the Zn sites to the O sites, indicating that the O sites occupy the ground state of the ZnO semiconductor. In other words, the Zn sites occupy the excited state of the ZnO semiconductor. In a perfect ZnO crystal, the Fermi level (E_F) is at the midline between E_{CBM} and E_{VBM}, as shown in figure 7.6(a). In a ZnO crystal with oxygen vacancies (V_O), the electrons in the Zn sites cannot be transferred to the V_O sites; they therefore increase the electron density in the conduction band of the ZnO crystal, which decreases the difference between E_{CBM} and E_F, as shown in figure 7.6(b). This explains why ZnO nanoparticles are always n-type semiconductor materials, owing to the presence of oxygen vacancies at the surface. In an AZO crystal, the Al dopants are used to replace Zn atoms. When an

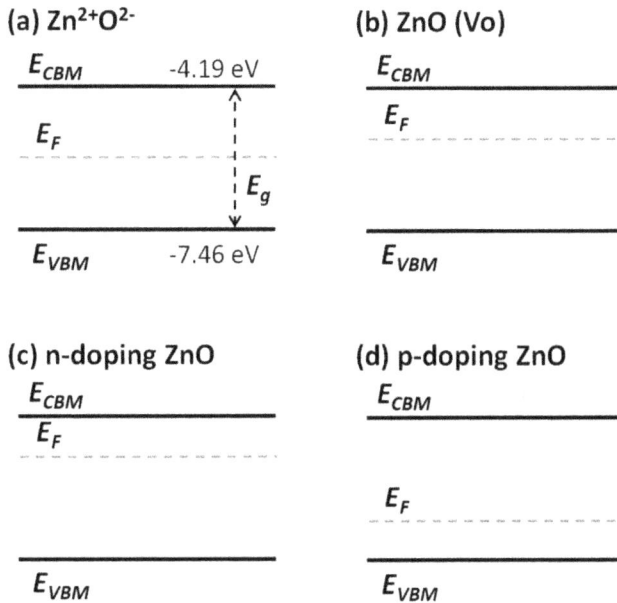

Figure 7.6. Energy diagrams. (a) ZnO. (b) ZnO with oxygen vacancies. (c) n-doped ZnO. (d) p-doped ZnO.

Al atom occupies a Zn site in the crystal structure, each Al atom can provide three electrons from its outer orbital to the adjacent oxygen sites, keeping one electron in the Al-occupied Zn site, which results in a smaller difference between E_{CBM} and E_F, as shown in figure 7.6(c). Gallium (Ga) atoms and/or indium (In) atoms can also be used to replace Al atoms as n-type dopants in ZnO semiconductors. When a Mg atom occupies a Zn site in the crystal structure, each Mg atom can provide one electron from its outer orbital to the adjacent O sites, thereby leaving one residual hole in the O sites, which results in a smaller difference between E_{VBM} and E_F, as shown in figure 7.6(d). On the other hand, p-type ZnO semiconductors can also be formed when nitrogen (N) atoms occupy the oxygen sites in the crystal structure [10]. When a N atom occupies an O site in the crystal structure, the N atom only accepts two electrons from the adjacent Zn atom, thereby leaving one residual hole in the N-occupied O site, which results in a smaller difference between E_{VBM} and E_F. The abovementioned concept can be used to distinguish the types of doping used in GaN and GaAs semiconductors. Zn and Mg can be used as p-type dopants in GaN and GaAs semiconductors. Si and Ge can be used as n-type dopants in GaN and GaAs semiconductors. GaN semiconductors have been used in applications such as light-emitting diodes and transistor devices, owing to their relatively higher exciton binding energy. GaAs semiconductors have been used in applications such as solar cells and photodetector devices, owing to their relatively lower exciton binding energy.

7.4 Carrier drift

In section 2.4, the Drude model was used to describe the energy transport mechanism of free carriers in conductive materials. In n-type and p-type semi-conductors, the energy transport mechanism of free carriers can also can be understood using the Drude model. Let us review the Drude model using the function of conductivity written in equation (7.3a), where σ is the conductivity, N is the free carrier density, τ is the relaxation time of the free carriers, and m^* is the effective mass of the free carriers. In the domain of Hall measurements, the conductivity function can be represented by equation (7.3b), where μ is the carrier drift mobility. In semiconductors, the carriers can be free electrons and/or free holes. Here, μ is used to represent the relationship between the applied electric field (E) and the carrier drift velocity (v_d), which is written in equation (7.3c). In the carrier drift process, the carrier relaxation time (τ) is defined as the mean time between collisions. The collisions are related to lattice vibration and ionized impurities. In other words, two relaxation pathways are available during the carrier drift process. When the collisions are dominated by lattice vibration, the lattice-vibration-mediated carrier mobility (μ_L) is related to the temperature (T) as shown in equation (7.3d). When the collisions are dominated by ionized impurity scattering, the impurity-mediated carrier mobility (μ_I) is related to the temperature and the ionized impurity density (N_I), as written in equation (7.3e). The net carrier mobility (μ_{net}) is written in equation (7.3f).

$$\sigma = e^2 N \tau / m^* \tag{7.3a}$$

$$\sigma = eN\mu \tag{7.3b}$$

$$v_d = \mu E \tag{7.3c}$$

$$\mu_L \propto 1/T^{3/2} \tag{7.3d}$$

$$\mu_I \propto 1/(N_I T^{3/2}) \tag{7.3e}$$

$$\frac{1}{\mu_{\text{net}}} = \frac{1}{\mu_L} + \frac{1}{\mu_I} \tag{7.3f}$$

$$\sigma_S = C_S eN/(N_I T^{3/2}) \tag{7.3g}$$

Equations (7.3d) and (7.3e) show that a higher temperature results in lower carrier mobility, which is due to the increase in the collision frequency between charge carriers (free carriers) and phonons (vibrating lattice). Equation (7.3e) shows that a higher ionized impurity density results in lower carrier mobility, which is due to an increase in the density of trap centers. Equation (7.3f) shows that μ_{net} is close to μ_I when μ_I is much lower than μ_L. In a heavily doped semiconductor, the function of conductivity can be written as equation (7.3g), where C_S is a coefficient related to crystallinity. In doped semiconductors, better crystallinity corresponds to larger values of C_S, thereby resulting in higher conductivity.

Equation (7.3c) shows that the carrier drift velocity (v_d) is linearly proportional to the applied electric field (E). However, there velocity saturation takes place due to the increase in collisions between the charge carriers and the vibrating lattice. The threshold carrier drift velocity (v_{th}) can be computed when the kinetic energy of the moving carriers equals the translational kinetic energy of the molecules in a crystal, as written in equation (7.4a), where m^* is the effective mass of the free carriers and $K_B T$ is the thermal energy. When the temperature is fixed, smaller values of m^* correspond to larger values of v_{th}, which means that smaller values of m^* result in more directional energy transport of free carriers. When m^* equals one electron mass, the calculated value of v_{th} is 1.17×10^5 m s^{-1}. The v_{th} value of a semiconductor can be determined by measuring its current–voltage (I–V) curve. Figure 7.7 displays the typical I–V curve of a metal/semiconductor/metal device when ohmic contacts are formed at the two interfaces. The v_{th} value of the semiconductor can be computed using equation (7.4b), where I_{sat} is the saturation current, A is the cross-sectional area of the semiconducting channel, e is the electron charge, and N is the free carrier density. By combining equations (7.4a) and (7.4b), the effective mass of the free carriers can be determined by analyzing the I–V curve when T, A, and N are known.

$$\frac{1}{2} m^* v_{\text{th}}^2 = \frac{3}{2} K_B T \tag{7.4a}$$

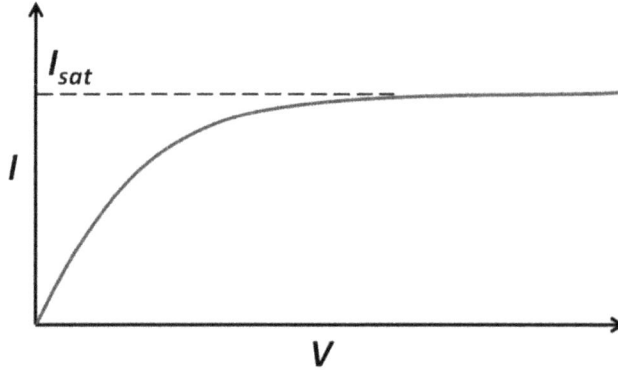

Figure 7.7. The current–voltage (I–V) curve of a metal/semiconductor/metal device with ohmic contacts.

$$\nu_{th} = I_{sat}/(AeN) \tag{7.4b}$$

In equation (7.3c), the carrier drift velocity (ν_d) is linearly proportional to the carrier mobility (μ). After combining equations (7.3a), (7.3b), and (7.3c), ν_d can be written as equation (7.5a). The longest propagation distance between two successive collisions equals $\tau\nu_{th}$, which is defined as the ballistic transport length in materials. The ballistic transport length (l) can be computed using equation (7.5b). In an AlGaN/GaN high electron mobility transistor (HEMT) device, the measured saturation drift velocity and carrier relaxation time are 3×10^5 m s^{-1} [11] and 1 ps [12], respectively, resulting in a long ballistic transport length of 500 nm. In other words, the energy loss of free carriers can be greatly reduced when the channel length of the HEMT device is shorter than the ballistic transport length.

$$\nu_d = \frac{e\tau}{m^*}E \tag{7.5a}$$

$$l = \nu_{th} \times \tau \tag{7.5b}$$

7.5 Carrier diffusion

Charge carriers can diffuse from a high-carrier density region to a low-carrier density region, thereby forming a current flow, which is an energy transport phenomenon. The relation between the diffusion current density ($J_{diffusion}$) and the carrier density gradient (dN_e/dx) is expressed in equation (7.6), where e is the electron charge and D_e is the electron diffusion coefficient. In a semiconductor bar, D_e can be determined by measuring the diffusion current when the electron density distribution is known. For example, the electron density varies linearly from 2×10^{18} cm^{-3} to 8×10^{18} cm^{-3} over a distance of 500 μm. When the measured $J_{diffusion}$ is 100 A cm^{-2}, the calculated D_e value is 5.2 cm^2 s^{-1}.

$$J_{diffusion} = eD_e\frac{dN_e(x)}{dx} \tag{7.6}$$

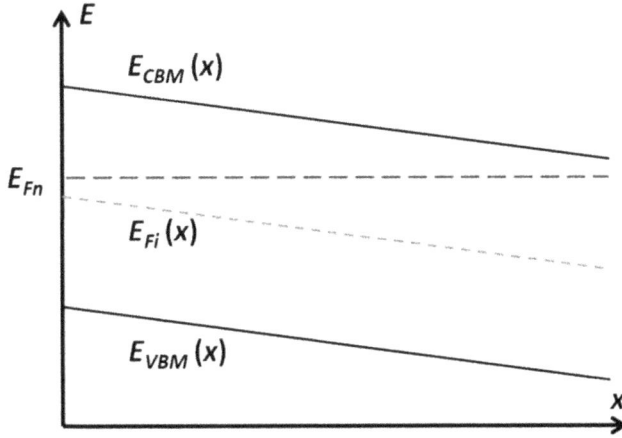

Figure 7.8. Space-dependent energy diagram of an n-type graded-dopant semiconductor.

The energy diagram for an n-type graded-dopant semiconductor is plotted in figure 7.8. E_{CBM}, E_{Fi}, and E_{VBM} vary linearly in real space (along the x-axis). The relation between the free electron density (N_e) and the Fermi level (E_{Fn}) is stated in equation (7.7a), where n_i is the intrinsic carrier concentration, E_{Fi} is the Fermi level in the intrinsic semiconductor, and $K_B T$ is the thermal energy. E_{Fi} is a function of N_e. Following rearrangement, the expression for E_{Fi} can be written as shown in equation (7.7b). The derivative of equation (7.7b) is given in equation (7.7c). In the space-dependent energy diagram, the E_{CBM} curve is parallel to the E_{Fi} curve, resulting in the relation $dE_{Fi}/dx = dE_{\text{CBM}}/dx$. In the conduction band, the energy of the free electrons is $e\phi_{\text{CBM}}$, where ϕ_{CBM} is the potential of the free electrons. Therefore, the graded potential ($d\phi_{\text{CBM}}/dx$) produces a built-in electric field (E_x) in the conduction band. In other words, the electrons are influenced by a built-in electric field in the conduction band of the n-type graded-dopant semiconductor that acts in the x-direction. The function used to calculate E_x is shown in equation (7.7d). According to Ohm's law, the E_x-induced current density can be written as $J_x = eN_e\mu_e E_x$, which is the drift current density, where e is the electron charge, N_e is the free electron density, and μ_e is the electron mobility. Following rearrangement, the E_x-induced drift current density can be expressed as equation (7.7e). In the steady state, the net current density ($J_{\text{diffusion}} + J_{\text{drift}}$) in the conduction band of the graded-dopant semiconductor is zero, resulting in the Einstein relation in semiconductors: $D_e = (K_B T/e)\mu_e$. In other words, the measured electron mobility can be used to compute the electron diffusion coefficient.

$$N_e = n_i \exp[(E_{Fn} - E_{Fi})/K_B T] \tag{7.7a}$$

$$-E_{Fi} = -E_{Fn} + K_B T \ln[N_e(x)/n_i] \tag{7.7b}$$

$$-\frac{dE_{Fi}}{dx} = \frac{K_B T}{N_e}\frac{dN_e(x)}{dx} \tag{7.7c}$$

$$E_x = -\frac{K_B T}{e N_e} \frac{dN_e(x)}{dx} \tag{7.7d}$$

$$J_{\text{drift}} = -\mu_e K_B T \frac{dN_e}{dx} \tag{7.7e}$$

In a packed molecular thin film, the energy transport mechanism of free carriers between adjacent molecules is similar to carrier diffusion, which can be described using equation (7.8a), where D is the carrier diffusion coefficient, a is the spacing between adjacent molecules, and R is the energy transfer rate from the excited molecules to the unexcited molecules. According to the Einstein relation in semiconductors, the relation between carrier drift mobility (μ) and D can be expressed as equation (7.8b). In other words, the carrier diffusion coefficient determines the carrier drift mobility in packed molecular thin films. The μ value of a packed molecular thin film can be measured experimentally and used to compute the D value. To evaluate the R value, the a value can be estimated by analyzing the x-ray diffraction (XRD) pattern using the Scherrer equation given in equation (7.8c), where K is a shape factor, λ is the wavelength of the x-ray source, β is the width of the XRD peak, and θ is the XRD peak angle. Conceptually, the spacings between adjacent molecules along the x, y, and z-axes are different, which is mainly related to the molecular structure. In other words, carrier diffusion is anisotropic in packed molecular thin films, which influences the application of the resultant organic thin films. When the D value is higher in the vertical direction (horizontal direction), the corresponding organic thin film is more suitable for use in solar cells/LEDs (field-effect transistors).

$$D = a^2 R \tag{7.8a}$$

$$\mu = \frac{e}{K_B T} D \tag{7.8b}$$

$$a = \frac{k\lambda}{\beta \cos \theta} \tag{7.8c}$$

Bibliography

[1] Kikuchi A 1999 Calculation of ohmic contact resistance at a metal/silicon interface *Phys. Stat. Sol.* A **175** 623–9
[2] Cheng L, Yang J-Y and Zheng. W 2022 Bandgap, mobility, dielectric constant, and baliga's figure of merit of 4H-SiC, GaN, β-Ga$_2$O$_3$ form 300 to 620 K *Appl. Electron. Matier.* **4** 4140–5
[3] Bae J W, Hossain T, Adesida I, Bogart K H, Koleske D, Allerman A A and Jang J H 2005 Low resistance ohmic contact to p-type GaN using Pd/Ir/Au multilayer scheme *J. Vac. Sci. Technol.* B **23** 1072
[4] Persson C, da Silva A F, Ahuja R and Johansson. B 2001 Effective electronic masses in wurtzite and zinc-blende GaN and AlN *J. Cyrstal Growth* **231** 397–406

[5] Huang L, Xia M and Gu. X 2020 A critical review of theory and progress in Ohmic contacts to p-type SiC *J. Cryst. Growth* **531** 125353

[6] Son N T, Hai P N, Chen W M, Hallin C, Monemar B and Janzen. E 2000 Hole effective masses in 4H SiC *Phys. Rev.* B **61** R10544–6

[7] Ponce S and Giustino F 2020 Structural, electronic, elastic, power, and transport properties of β-Ga_2O_3 from first principles *Phys. Rev. Res.* **2** 033102

[8] Khan M A M, Kumar S, Khan M N, Ahamed M and Al Dwayyan. A S 2014 Microstructure and blueshift in optical band gap of nanocrystalline $Al_xZn_{1-x}O$ thin films *J. Lumin.* **155** 275–18

[9] Wu H-C, Peng Y-C and Shen T-P 2012 Electronic and optical properties of substitutional and interstitial Si-doped ZnO *Materials* **5** 2088–100

[10] Chen J Y, Zhang H T, Chen Q, Hsuian F and Cherng. J-S 2020 Stable p-type nitrogen-doped zinc oxide films prepared by magnetron sputtering *Vacuum* **180** 109576

[11] Ardaravicius L, Matulionis A, Liberis J, Kiprijanovic O, Ramonas M, Eastman L F, Shealy J R and Vertiatchikh A 2003 Electron drift velocity in AlGaN/GaN channel at high elevtric fields *Appl. Phys. Lett.* **83** 4038–40

[12] Nafari M, Aizin G R and Jornet. J M 2018 Plasmonic HEMT terahertz transmitter based on the Dyakonov-Shur instability: performance analysis and impact of nonideal boundaries *Phys. Rev. Appl.* **10** 064025

IOP Publishing

Light–Material Interactions and Applications in
Optoelectronic Devices

Anjali Chandel and Sheng Hsiung Chang

Chapter 8

Quantum wells and quantum dots

In this chapter, the quantum-scale effects of metals and semiconductors are briefly described to demonstrate how physical size can change the properties of materials. In the section on quantum wells (QWs), a formula is used to discuss electron–hole coulomb interactions, which include carrier confinement in the well region and wave distribution in the barrier region. The lattice constant and thermal expansion coefficient mismatches between AlN (GaN) and GaN (InN) are discussed to illustrate the design rules of QW-based light emission devices. In the section on quantum dots (QDs), progress in the development of InAs QDs is used to realize a high-performance QD-based light emission device. It should be noted that InAs QDs show great potential for use in light-emitting diode devices in the near-infrared. Wavelengths ranging from about 1500 nm to about 3000 nm can be achieved by changing the diameter of the QDs from 5 to 15 nm.

8.1 Quantum-scale effects

In crystal materials, the manipulation of effective carrier masses is known as a quantum-scale effect; such changes can greatly change the optical, electrical, and optoelectronic properties of semiconductors. However, the effective masses of free electrons in metals are insensitive to physical size. It is known that the bound electrons in semiconductors (the free electrons in metals) are localized (delocalized); as a result, they are sensitive (insensitive) to structure-induced changes in the periodic potential in the crystal. In other words, the size-dependent optical, electrical, and optoelectronic properties are apparent characteristics which can be used to evaluate the interaction strength between electrons and the crystal structure, which is a quantum-scale effect.

doi:10.1088/978-0-7503-6099-9ch8   8-1   © IOP Publishing Ltd 2024. All rights,

8.2 Quantum wells

In planar optoelectronic devices, an alloyed tri-layered semiconductor heterostructure can be used to increase the localization of charged carriers (electrons and/or holes) owing to the formation of electrical barriers for electrons and holes, as shown in figure 8.1(a), where D is the thickness of the GaN layer, ε_b is the optical dielectric constant of AlGaN, and ε_w is the optical constant of GaN. Spatially separated electron and hole clouds are illustrated in figure 8.1(b), where (x_e, y_e, z_e) and (x_h, y_h, z_h) are their central points, respectively. The potential energy of the electron–hole coulomb interaction of the QW can be computed using equation (8.1a), where ρ $(=\sqrt{(x_e - x_h)^2 + (y_e - y_h)^2})$ is the distance between the electron and hole clouds in the x–y-plane, δ_ε is related to the permittivities of AlGaN and GaN, and B_1 and B_2 are related to D, z_e, z_h, and δ_ε [1]. δ_ε, B_1, and B_2 can be computed using equations (8.1b), (8.1c), and (8.1d), respectively. J_0 is a Bessel function of the first kind when the index is zero.

Let us discuss the effects of D, ε_w, and ε_b on V_{eh}. When the D value is infinity, the B_1 and B_2 values are zero, which results in the anti-confinement effect. When the ε_w and ε_b values are the same, δ_ε is zero, which also results in the anti-confinement effect. In other words, a smaller D value and/or a larger δ_ε value can result in a higher V_{eh} value, thereby increasing the exciton binding energy of the alloyed tri-layered semiconductor heterostructure. It is surprising that the V_{eh} value approaches infinity when the D value and the δ_ε value are close to zero and infinity, respectively. In other words, it is a perfect QW when the δ_ε value is infinite.

$$|V_{eh}| = \frac{e^2}{4\pi\varepsilon_0\varepsilon_w}\left[\frac{1}{(z_e - z_h)^2 + \rho^2} + 2\delta_\varepsilon B_1 + 2\delta_\varepsilon^2 B_2\right] \qquad (8.1a)$$

$$\delta_\varepsilon = (\varepsilon_w - \varepsilon_b)/(\varepsilon_w + \varepsilon_b) \qquad (8.1b)$$

$$B_1 = \int_0^\infty \frac{\exp(-\eta D)\cosh[\eta(z_e + z_h)]}{1 - \delta_\varepsilon^2 \exp(-2\eta D)} J_0(\eta\rho)d\eta \qquad (8.1c)$$

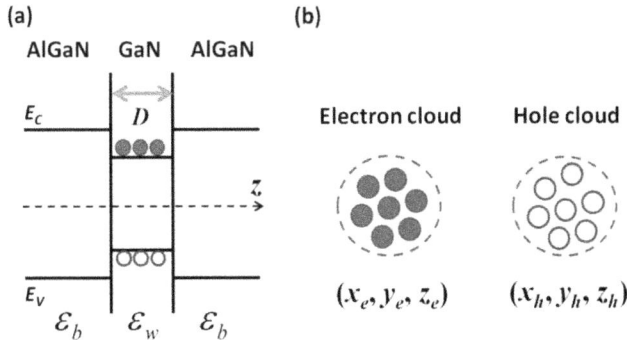

Figure 8.1. (a) Energy diagram of a tri-layered alloyed AlGaN/GaN/AlGaN structure. (b) Spatially separated electron and hole clouds.

$$B_2 = \int_0^\infty \frac{\exp(-2\eta D)\cosh[\eta(z_e - z_h)]}{1 - \delta_\varepsilon^2 \exp(-2\eta D)} J_0(\eta\rho)d\eta \qquad (8.1d)$$

In an infinite QW, the energy states and wave functions of the electrons can be written as equations (8.2a) and (8.2b), respectively, where $\hbar$ is the reduced Planck constant, m_e^* is the effective electron mass, D is the thickness of the QW, and n is the quantum number, which is a positive integer [2]. When m_e^* is independent of the quantum number, larger values of n correspond to higher values of E_n. Therefore, E_2 (E_3) is higher than E_1 (E_2). The penetration of the electron waves into the infinite barriers is zero, which indicates that the values of the electron wave functions at the boundaries are zero. Therefore, the electron waves form resonant modes with sinusoidal functions, as shown in figure 8.2. Conceptually, the relaxation time of the electrons decreases as the quantum number increases owing to the probability of charge carrier–lattice interaction between the two barriers. The quantum-number-dependent relaxation time (τ_n) of the QW can be computed using equation (8.2c), where ε_0 is the absolute permittivity of vacuum and f_n is the oscillation strength. The τ_n value decreases as the the E_n value increases. In other words, the quantum confinement effects result in faster carrier dynamics. It should be noted that τ_n is proportional to D^4 when e_w, f_n, and m_e^* are independent of physical size.

$$E_n = \frac{\hbar^2}{2m_e^*}\left(\frac{n\pi}{D}\right)^2 \qquad (8.2a)$$

$$\phi_n = A_n \sin(n\pi z/D) \qquad (8.2b)$$

$$\tau_n = \frac{2\pi\varepsilon_0 m_e^* c^3 \hbar^2}{\sqrt{\varepsilon_w}\, e^2 E_n^2 f_n} \qquad (8.2c)$$

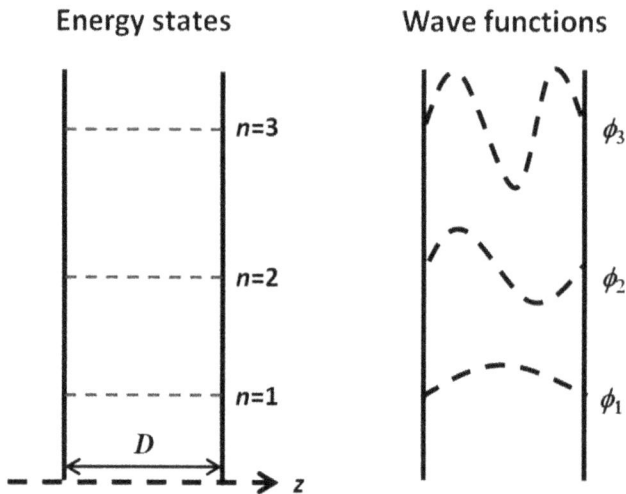

Figure 8.2. Energy states and wave functions of bound electrons in a perfect QW.

In a finite QW, the higher-order modes are suppressed when the barrier height is decreased to a lower level than their resonant energies. On the other hand, the electron waves penetrate the potential barrier layers, as shown in figure 8.3. The reduced barrier height results in weaker confinement effects owing to the longer penetration depth of the electron waves, which means that the exciton binding energy (carrier confinement) is proportional to the barrier height of the QW. According to equation (8.2a), narrower D values result in larger E_n values, thereby suppressing the existence of higher-order modes in the QW.

$Al_xGa_{1-x}N/GaN/Al_xGa_{1-x}N$ and $GaN/In_xGa_{1-x}N/GaN$ are widely used QW materials that can efficiently generate blue light and green light, respectively [3, 4]. When the x value equals unity, the QW materials become AlN/GaN/AlN and GaN/InN/GaN, respectively. Electrons and holes can be confined in the GaN layer of AlN/GaN/AlN and InGaN layer of layered GaN/InGaN/GaN because the E_g value of AlN (GaN) is larger than that of GaN (InGaN). The bandgap, lattice constant, and thermal expansion coefficient values of AlN, GaN, and InN are listed in table 8.1. However, the lattice constants and thermal expansion coefficients of AlN, GaN, and InN are not perfectly matched, which can result in strain effects in QW-based light-emitting devices due to residual stress. The lattice mismatch (LM) can be computed using the simple relation: $LM = (a_s - a_f)/a_s$, where a_s and a_f are the lattice constant values of the substrate and the deposited film, respectively. When AlN

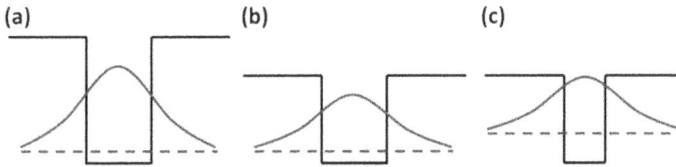

Figure 8.3. Electron waves in finite QWs with different barrier heights. (a) Higher barrier height. (b) Lower barrier height. (c) Lower barrier height and narrower well width.

Table 8.1. Bandgaps, lattice constants, and thermal expansion coefficients of wurtzite-phase AlN, GaN, and InN at $T = 300$ K.

Compound	E_g (eV)	Lattice constant (nm)	Thermal expansion coefficient ($\times 10^{-6}$ K^{-1})	References
AlN	6.20	$a = b = 0.311$ $c = 0.489$	$\Delta a/a = 4.20$ $\Delta c/c = 5.30$	[7–9]
GaN	3.44	$a = b = 0.319$ $c = 0.518$	$\Delta a/a = 5.59$ $\Delta c/c = 3.17$	[8, 10, 11]
InN	0.63	$a = b = 0.354$ $c = 0.571$	$\Delta a/a = 3.80$ $\Delta c/c = 2.90$	[8, 12, 13]

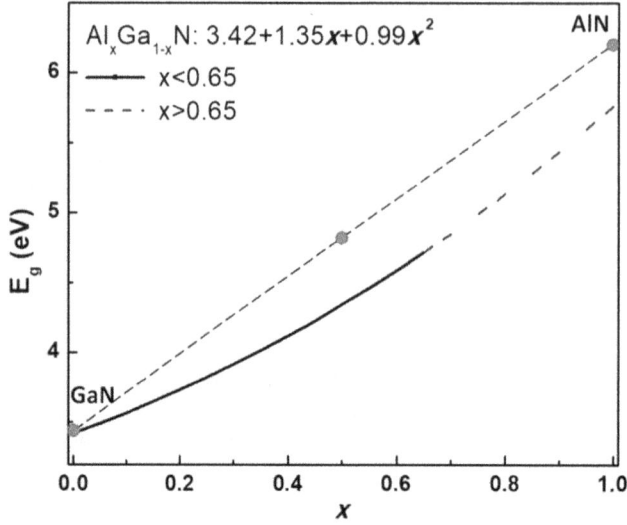

Figure 8.4. x–E_g curve of wurtzite-phase $Al_xGa_{1-x}N$ and the E_g values of GaN and AlN at $T = 300$ K.

(GaN) and GaN (InN) are the substrate and the film, respectively, the calculated LM value is -2.50% (-9.89%). To reduce the lattice mismatch between AlN (GaN) and GaN (InN), the x value is adjusted in the $Al_xGa_{1-x}N$ ($In_xGa_{1-x}N$) alloy semiconductor, which is used as a buffer layer. At $T = 300$ K, the x-dependent E_g value of $Al_xGa_{1-x}N$ can be computed using equation (8.3) for values of x lower than 0.65 [5]. The calculated E_g value of $Al_{0.5}Ga_{0.5}N$ is about 4.34 eV, which is lower than the average value of 4.82 eV for the E_g values of AlN and GaN in table 8.1. The x–E_g curve and the E_g values of AlN, $Al_{0.5}Ga_{0.5}N$, and GaN are plotted in figure 8.4. It should be noted that the black solid line was determined from experimental results. It describes a quadratic curve when the x value is lower than 0.65, but when the x value ranges from 0.65 to 1, the blue dashed line in figure 8.4 cannot be used to correctly predict the E_g value of an $Al_xGa_{1-x}N$ alloy semiconductor, which has been explained as being due to the decrease in crystallinity as the x value increases. In $Al_xGa_{1-x}N$ alloy semiconductors, the formation of AlN increases as the x value increases, thereby decreasing the crystallinity of the resultant $Al_xGa_{1-x}N$. It should be noted that less crystalline semiconductors result in a more distorted energy distribution in the QW, thereby broadening the quantized energy states, which can be observed in the absorbance spectra [6].

$$E_g(x) = 3.42 + 1.35x + 0.99x^2 \tag{8.3}$$

8.3 Quantum dots

In the previous section, QWs were shown to localize electrons and holes, resulting in higher exciton binding energy and thereby increasing light emission efficiency. However, it is difficult to form highly crystalline and strongly confined QWs. Conceptually, it is possible to form QDs in a QW system using the lattice mismatch between the deposited film and the substrate. InAs QDs have been fabricated on top of GaAs substrates via the Stranski–Krastanow transition during the cooling processes of the metal–organic chemical vapor deposition (MOCVD) or molecular beam epitaxy (MBE) methods [14, 15]. These QDs which can be used over a wide spectral range from about 700 to about 1500 nm [16, 17]. Self-assembled InAs QDs have pyramidal shapes due to the lattice mismatch that occurs during the dewetting process associated with cooling, which results in difficulties in analytically resolving the quantized energy levels of the pyramid-shaped QDs. Fortunately, the quantized energy levels of the pyramid-shaped QDs can be solved numerically [18].

When electrons and holes are three-dimensionally confined in a semiconductor QD, the exciton binding energy can increase greatly from several meV to several eV due to a decrease in the physical size of the QD. However, the surface defects of QDs result in additional non-radiative decay, thereby decreasing light emission efficiency. Therefore, effective surface passivation of QDs is extremely important to realize highly efficient QD-based optoelectronic devices. To passivate the surface defects of InAs QDs, larger-bandgap semiconductors such as CdS, CdSe, ZnS, and ZnSe are widely used as capping materials. However, the low crystallinity of the capping materials intrinsically results in low light emission efficiency. The low crystallinity of the capping material is mainly due to the lattice mismatch between the QD and the capping material. Even though the crystallinity of the capping material can be improved, the surface defects of the thin capping materials still result in additional non-radiative decay. Therefore, the capping materials must be thick in order to minimize the negative influence on the optoelectronic properties of the QDs. Fortunately, ligand passivation can be used to effectively suppress the defect-mediated non-radiative decay of the QDs, thereby increasing the light emission efficiency [19]. The three-dimensional (3D) molecular structures of the widely used carboxylic acid ligands are plotted in figure 8.5. In figures 8.5(a)–(c), the use of carbon ring structures and carbon double bonds results in planar structures. After the dehydrogenation of the carboxyl (–COOH) group, these planar organic ligands can be conformally loaded on the surface of QDs, effectively passivating the surface defects.

The bandgap values of small organic molecules are about 4 eV, which can be used to confine excitons in low-bandgap QDs. The lowest unoccupied molecular orbital

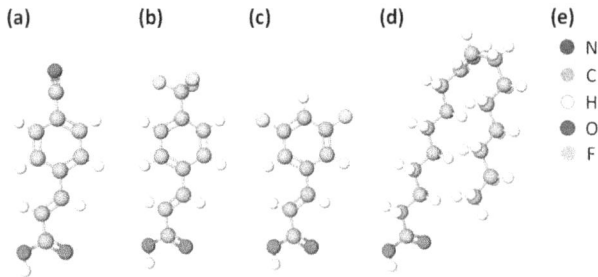

Figure 8.5. 3D molecular structures of carboxylic acid ligands. Blue, gray, white, red, and green spheres are nitrogen, carbon, hydrogen, oxygen, and fluoride atoms, respectively.

(LUMO) values of carboxyl-based organic molecules are about -3.5 eV, which is higher than the conduction band minimum energy level (-5.15 eV) of bulk InAs. This is shown in the energy diagram of a carboxylic acid ligand-capped InAs QD plotted in figure 8.6. The bandgap of the QD can be experimentally determined from the absorption tail of the lowest exciton transition peak, which is the lowest transition energy of electrons from the ground state to the excited state. The lowest transition energy of electrons in the QD is the sum of the bandgap energy, the quantum confinement energy, and the negative exciton binding energy, which can be written as equation (8.4a), where E_g^{QD} is the QD's bandgap energy, E_g is the bulk's bandgap energy, E_{QC} is the quantum confinement energy, and E_{Ry} is the exciton binding energy (Rydberg energy). The definition of Rydberg energy is given in section 6.2. The E_{QC} and E_{Ry} values are both related to the effective properties of electrons and holes, which can be computed using equations (8.4b) and (6.3b), respectively, where a is the radius of the QD, m_e^* is the effective mass of an electron, and m_h^* is the effective mass of a hole.

$$E_g^{QD} = E_g + E_{QC} - E_{Ry} \qquad (8.4a)$$

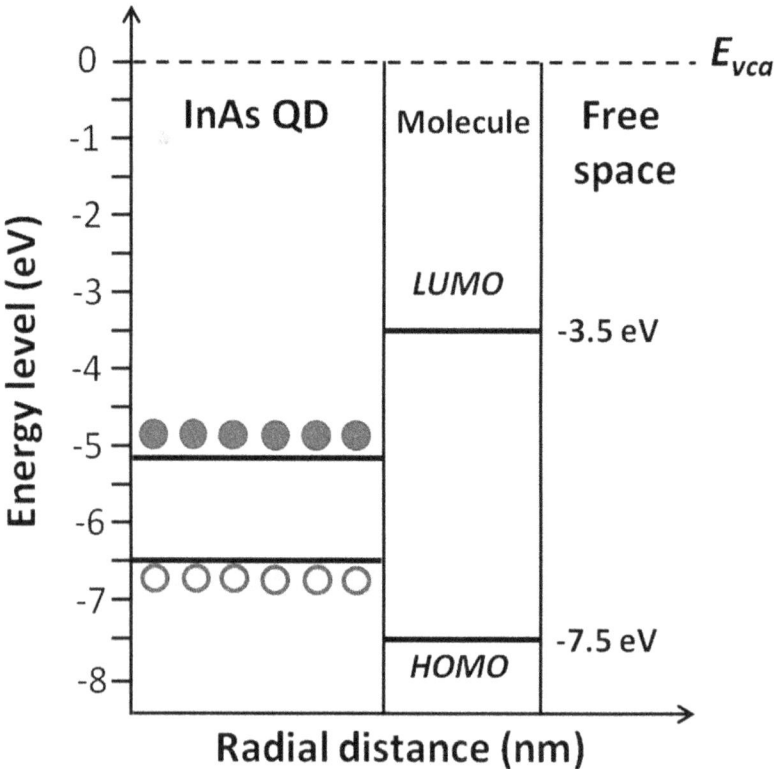

Figure 8.6. Energy diagram of an InAs QD with a capping layer of carboxylic acid ligands.

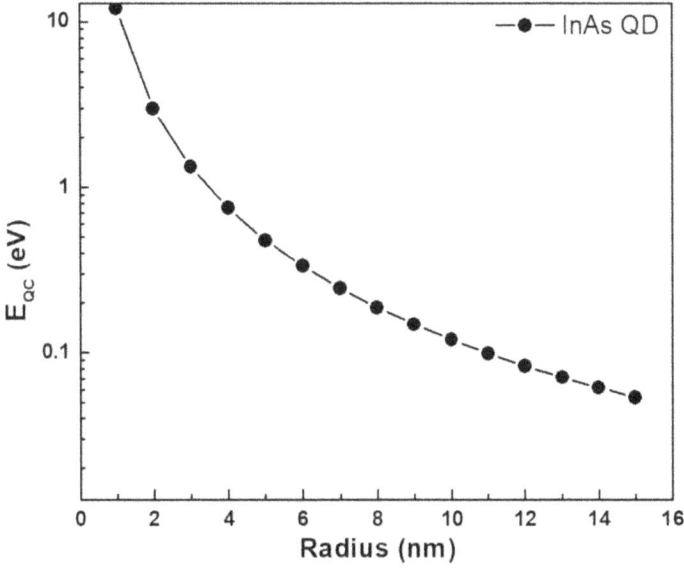

Figure 8.7. Quantum confinement energy (E_{QC}) of a perfect InAs QD for radius values ranging from 1 to 15 nm.

$$E_{QC} = \frac{\hbar^2 \pi^2}{2a^2}\left(\frac{1}{m_e^*} + \frac{1}{m_h^*}\right) \qquad (8.4b)$$

Let us compute the E_{QC} of a perfect InAs QD. In bulk InAs, the effective masses of electrons and holes are 0.034 and 0.4 m_e, respectively [20]. When the effective masses of charge carriers are independent of the physical size, E_{QC} is inversely proportional to a^2, as shown in figure 8.7. The E_{QC} value greatly increases from 53.3 to 479.6 meV as the radius of the InAs QD increases from 15 to 5 nm. The large change in the E_{QC} of the InAs QD can be explained as being due to the large Bohr radius of 34 nm [21]. These E_{QC} values are far higher than the E_{Ry} value of bulk InAs (4 meV). In other words, the change in E_g^{QD} is mainly caused by the change in E_{QC}. The E_g and E_{Ry} values of bulk InAs are 0.36 eV and 4 meV [22, 23], respectively, which results in a decrease in the bandgap wavelength from 3020 to 1483 nm as the radius of the InAs QD decreases from 15 to 5 nm. The ultrawide tunability of the bandgap wavelength demonstrates the potential of InAs QDs for use in optoelectronic devices in the near-infrared wavelength range. It should be noted that the prediction accuracy of E_{QC} decreases as the radius of the QDs decreases because the effective masses of electrons and holes increase as the physical size decreases at the nanoscale [24].

Bibliography

[1] Gerlach B and Wusthoff J 1998 Exciton binding energy in a quantum well *Phys. Rev.* B **58** 10568–77

[2] Fox M and Ispasoiu R 2006 Quantum wells, superlattices, and band-gap engineering *Springer Handbook of Electronic and Photonic Materials* (New York: Springer) 1020–40

[3] Iwaya M, Terao S, Sano T, Takanami S, Ukai T, Nakamura R, Kamiyama S, Amano H and Akasaki I 2001 High-efficiency GaN/Al$_x$Ga$_{1-x}$N multi-quantum-well light emitter grown on low-dislocation density Al$_x$Ga$_{1-x}$N *Phys. Status Solidi (A)* **188** 117–20

[4] Tzou A-J *et al* 2016 High-performance InGaN-based green light-emitting diodes with quaternary InAlGaN/GaN superlattice electron blocking layer *Opt. Express* **24** 11387–95

[5] Tisch U, Meyler B, Katz O, Finkman E and Salzman J 2001 Dependence of the refractive index of Al$_x$Ga$_{1-x}$N on temperature and composition at elevated temperatures *J. Appl. Phys.* **89** 2676–85

[6] Podlipskas Z, Aleksiejunas R, Kadys A, Mickevicius J, Jurkevicius J, Tamulaitis G, Shur M, Shatalov M, Yang J and Gaska R 2016 Dependence of radiative and nonradiative recombination on carrier density and Al content in thick AlGaN epilayers *J. Phys.* D **49** 145110

[7] Lu Z, Wang F and Liu Y 2021 The first principle calculation of improving p-type characteristics of B$_x$Al$_{1-x}$N *Sci. Rep.* **11** 12720

[8] Araujo R B, de Almeida J S and da Silva A F 2013 Electronic properties of III-nitride semiconductors: a first-principles investigation using the Tran-Blaha modified Becke–Johnson potential *J. Appl. Phys.* **114** 183702

[9] Yim W M and Paff R J 1974 Thermal expansion of AlN, sapphire, and silicon *J. Appl. Phys.* **45** 1456–7

[10] Monemar B 1974 Fundamental energy gap of GaN from photoluminescence excitation spectra *Phys. Rev.* B **10** 676

[11] Gian W, Skowronski M and Rohrer G S 1996 Structural defects and their relationship to nucleation of GaN thin films *MRS Proc.* **423** 475–86

[12] Ishitani Y, Masuyama H, Terashima W, Yoshitani M, Hashimoto N, Che S B and Yoshikawa A 2005 Bandgap energy of InN and its temperature dependence *Phys. Status Solidi (c)* **2** 2276–80

[13] Zubrilov A 2001 *Properties of Advanced Semiconductor Materials: GaN, AlN, InN, BN, SiC, SiGe* ed M E Levinshtein, S L Rumynatsev and M S Shur (New York: Wiley) pp 49–66

[14] Li S *et al* 2013 InAs/GaAs quantum dots with wide-range tunable densities by simply varying V/III ratio using metal–organic chemical vapor deposition *Nanoscale Res. Lett.* **8** 367

[15] Kwoen J, Jang B, Lee J, Kageyama T, Watanabe K and Arakawa Y 2018 All MBE grown InAs/GaAs quantum dot lasers on on-axis Si (001) *Opt. Express* **26** 11568–76

[16] Aharoni A, Mokari T, Popov I and Banin U 2006 Synthesis of InAs/CdSe/ZnSe core/shell1/shell2 structures with bright and stable near-infrared fluorescence *J. Am. Chem. Soc.* **128** 257–64

[17] Zimmer J P, Kim S-W, Ohnishi S, Tanaka E, Frangioni J V and Bawendi M G 2006 Size series of small indium arsenide-zinc selenide core–shell nanocrystals and their application to *in vivo* imaging *J. Am. Chem. Soc.* **128** 2526–7

[18] Hwang T-M, Lin W-W, Wang W-C and Wang W 2004 Numerical simulation of three dimensional pyramid quantum dot *J. Comput. Phys.* **196** 208–32

[19] Kroupa D M, Voros M, Brawand N P, Mcnichols B W, Miller E M, Gu J, Nozik A J, Sellinger A, Galli G and Beard M C 2017 Tuning colloidal quantum dot band edge positions through solution-phase surface chemistry modification *Nat. Commun.* **8** 15257

[20] Bouarissa N and Aourag H 1999 Effective masses of electrons and heavy holes in InAs, InSb, GaSb, GaAs and some of their ternary compounds *Infrared Phys. Technol.* **40** 343–9

[21] Efros A L and Rosen M 2000 The electronic structure of semiconductor nanocrystals *Ann. Rev. Mater. Res.* **30** 475–521

[22] Bhat R, Dutta P S and Guha S 2008 Crystal growth and below-bandgap optical absorption studies in InAs for non-linear optic applications *J. Cryst. Growth* **310** 1910–6

[23] de-Leon S and Laikhtman B 2000 Exciton wave function, binding energy, and lifetime in InAs/GaSb coupled quantum wells *Phys. Rev.* B **61** 2874

[24] Bekhouche H, Rahou D, Gueddim A, Abdelhafidi M K and Bouarissa N 2018 Electron states, effective masses and transverse effective charge of InAs quantum dots *Opt. Quantum Electron.* **50** 309

IOP Publishing

Light–Material Interactions and Applications in
Optoelectronic Devices

Anjali Chandel and Sheng Hsiung Chang

Chapter 9

Photovoltaic solar cells

In this chapter, single-junction solar cells are classified using the relationship between the exciton binding energy (E_b) of their light-absorbing material and their thermal energy ($K_B T$). When the E_b value of the light-absorbing material is lower than $K_B T$, the device architecture of the resultant solar cell is a multilayered structure. When the E_b value of the light-absorbing material is higher than $K_B T$, the light-absorbing material must be mixed with an n-type semiconducting material in order to effectively perform exciton dissociation. In addition, the development progress and working mechanisms of Si, GaAs, InP, CdTe, copper indium gallium selenide (CIGS), copper zinc tin sulfide selenium (CZTSSe), and methylammonium lead triiodide (MAPbI3) solar cells are briefly described. In the last section, the device performance of solar cells is evaluated by analyzing current density–voltage curves under one sun of illumination, which is used to compute the photovoltaic performance, such as the open-circuit voltage, short-circuit current density, fill factor, and current hysteresis.

9.1 Classification of single-junction solar cells

Single-junction photovoltaic solar cells can be classified into inorganic solar cells and organic solar cells, mainly depending on the type of light-absorbing material used. Si, GaAs, InP, CdTe, CIGS, and CZTSSe are widely used inorganic light-absorbing materials. Poly(3-hexylthiophene):1-3(methoxycarbonyl)propyl-1phenyl [6,6]C_{61} (P3HT:PCBM) and poly[(2,6-(4,8-bis(5-(2-ethylhexyl-3-fluoro)thiophen-2-yl)-benzo[1,2-b:4,5-b′]dithiophene))-alt-(5,5-(1′,3′-di-2-thienyl-5′,7′-bis(2-ethylhexyl) benzo[1′,2′-c:4′,5′-c′]dithiophene-4,8-dione)]:2,2′-[[12,13-bis(2-ethylhexyl)-12,13-dihydro-3,9-diundecylbisthieno[2′,3′:4′,5′]thieno[2′,3′:4,5]pyrrolo[3,2-e:2′,3′-g][1–3] benzothiadiazole-2,10-diyl]bis[methylidyne(5,6-difluoro-3-oxo-1H-indene-2,1(3H)-diylidene)]]bis[propanedinitrile] (PM6:Y6) blended thin films are widely used organic light-absorbing materials. In addition, perovskite and quantum dot (QD) thin films have been shown to possess high photovoltaic responses.

The device structures of solar cells are mainly related to the light absorption coefficient and exciton diffusion length of the light-absorbing material used. In chapter 6, the relationship between the light absorption coefficient and exciton binding energy (exciton diffusion length) was examined, which can be used to understand the device structures used in highly efficient solar cells. In other words, the properties of the materials used to make solar cells can reasonably be used to classify them into different types. When the exciton binding energy is lower than the thermal energy $(K_B T)$ at room temperature, excitons can self-dissociate to form free electrons and free holes in the light-absorbing layer, thereby generating photo-current. When the exciton binding energy is higher than the thermal energy $(K_B T)$ at room temperature, excitons must diffuse to the interfaces in order to form free electrons and free holes in the electron transport material and the hole transport material, respectively, which indicates that the exciton diffusion length must be longer than the light absorption length in order to effectively generate the photo-current. The light absorption coefficient, exciton binding energy, and carrier diffusion length of some widely used light-absorbing materials are listed in table 9.1. Si, GaAs, InP, CdTe, CIGS, CZTSSe, and $MAPbI_3$ are three-dimensional (3D) crystal semiconductor materials, which have relatively lower E_b values and/or relatively longer carrier diffusion lengths, thereby forming efficient solar cells with planar multilayer structures. P3HT and PM6 are a conjugated polymer and a conjugated molecule, respectively, which have larger E_b values (308 and 140 meV), thereby forming efficient solar cells with bulk heterojunction (HJ) active layer structures.

It should be noted that Si has a weak light absorption coefficient ($\sim 0.1 \times 10^4$ cm^{-1}), which explains the thick Si layer used in efficient Si solar cells. To efficiently absorb sunlight, the thickness of the Si layer should be about 100 μm. The other 3D crystal semiconductor materials have stronger light absorption abilities, resulting in

Table 9.1. Absorption coefficient (α), exciton binding energy (E_b), and carrier diffusion length (L_D) values of some widely used light-absorbing materials in various types of solar cells.

Material	α (1 cm^{-1}) at $\lambda = 500$ nm	E_b (meV)	L_D (nm)	Type	References
Si	$\sim 0.10 \times 10^4$	~ 15	None	Planar	[1, 2]
GaAs	$\sim 1.07 \times 10^5$	~ 4.2	None	Planer	[3, 4]
InP	$\sim 1.26 \times 10^5$	~ 0.4	None	Planar	[5, 6]
CdTe	$\sim 1.07 \times 10^5$	~ 15	~ 6000	Planar	[7–9]
CIGS	$\sim 1.40 \times 10^5$	None	~ 3600	Planar	[10, 11]
CZTSSe	$\sim 1.50 \times 10^5$	None	~ 1200	Planar	[11, 12]
$MAPbI_3$	$\sim 2.01 \times 10^5$	~ 7	~ 6000	Planar	[13–15]
P3HT:PCBM	$\sim 2.00 \times 10^5$	~ 308	~ 125	Bulk HJ	[16–18]
PM6:Y6	$\sim 0.90 \times 10^5$	~ 140	~ 160	Bulk HJ	[19–21]

a thinner active layer in the resultant solar cells. In efficient solar cells, the required thicknesses of these compound semiconductor materials can be less than 1 μm, which is shorter than their carrier diffusion length values. The E_B values of P3HT in P3HT:PCBM film and PM6 in PM6:Y6 films are significantly larger than their thermal energies at room temperature, preventing the effective formation of free carriers. In P3HT:PCBM blended thin film, the diffusion length of excitons in the P3HT polymers is about 7 nm, which is far shorter than the required thickness of the light-absorbing material. Therefore, light-absorbing organic materials (P3HT or PM6) and n-type organic materials (PCBM or Y6) are blended to obtain longer carrier diffusion lengths (125 and 160 nm) that are close to the required thicknesses of the light-absorbing materials. In binary blended thin films, the vertical and horizontal distributions of the organic materials play an important role in efficient organic solar cells. In table 9.1, the optoelectronic properties of dyes are not listed owing to the limited information available and their rare discussion in the literature. The performance of the two organic blended thin films is determined by the relation between their light absorption coefficient, exciton binding energy, and exciton diffusion length.

Si, GaAs, InP, CdTe, CIGS, CZTSSe, and MAPbI$_3$ can be used as light absorbers in planar multilayer-structure-based solar cells. P3HT:PCBM and PM6:Y6 blended thin films can be used in bulk HJ solar cells. Moreover, the active layer of dye-sensitized solar cells (DSSCs) is similar to the active layer of bulk HJ solar cells. In DSSCs, the surface of mesoporous TiO$_2$ nanoparticles is sensitized with dyes to effectively collect the photogenerated electrons from the dyes in the TiO$_2$ nanoparticles.

9.2 Si solar cells

Crystalline silicon (c-Si) is an indirect, narrow-bandgap semiconductor material. Its diamond cubic crystal structure, with a lattice constant of 0.543 07 nm and an atomic number of 14, can be used to compute the density of c-Si, which is about 2.3290 g cm^{-3}. The indirect bandgap energy of crystalline Si is 1.12 eV, which corresponds to a bandgap wavelength of 1107 nm. When the absorption bandgap energy is 1.12 eV, the maximum current density of the Si solar cells is about 45 mA cm^{-2}. At $\lambda = 970$ nm, the absorption coefficient of c-Si is about 100 cm^{-1}. According to the Beer–Lambert law, c-Si can absorb about 90% of the incident photons at $\lambda = 970$ nm when its thickness is 230 μm, which explains the typical thickness of the silicon wafers used in solar cells.

To effectively absorb sunlight and collect photogenerated current, the dual surfaces of the c-Si substrate must be modified simultaneously. The top surface of the c-Si substrate must be textured to reduce reflection at the air/c-Si interface. When the incident angle is zero, the reflectance of the air/c-Si interface is about 38.8% at $\lambda = 500$ nm. In addition, the top and bottom surfaces of the c-Si substrate must be heavily doped to form n+ and p+ c-Si contacts, respectively. Figure 9.1 displays a cross-sectional schematic view of a passivated emitter, rear locally diffused (PERL) c-Si solar cell [22]. The finger electrode and rear electrode are used to collect

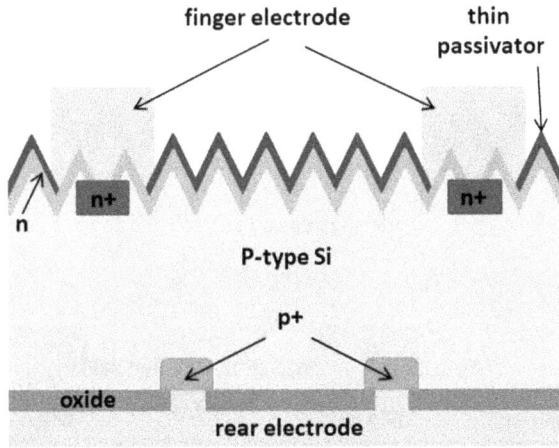

Figure 9.1. Cross-sectional schematic view of a PERL c-Si solar cell.

photogenerated electrons and holes, respectively. The textured Si surfaces can be made via a wet etching process [23], which can reduce the reflectance of the Si substrate. To further decrease the reflectance, a double layer can be deposited on top of the textured Si surface as an antireflective (AR) coating. The design method of an AR coating layer is mentioned in chapter 2. Phosphorus can be used as a Si dopant with a tunable concentration of up to 3×10^{20} cm^{-3}, which results in n-type Si or n+ type Si [24, 25]. Boron can be used as a Si dopant at a high concentration of 2×10^{19} cm^{-3}, which results in p+ type Si [26]. A thin passivator of n-type Si can be made from SiN$_x$, which can be fabricated using the plasma-enhanced chemical vapor deposition (PECVD) method [27]. PO$_x$/Al$_2$O$_3$ multilayers can be used to effectively passivate n-type textured Si surfaces by reducing the surface recombination velocity of free electrons, which can be decreased to 2 cm s^{-1} [28]. The oxide can be made from multilayered SiO$_2$/Al$_2$O$_3$, which can effectively passivate the p-type Si surface and thereby reduce positive charge accumulation in the bottom region of p-type Si. SiO$_2$/Al$_2$O$_3$ multilayers can be fabricated using the atomic layer deposition (ALD) method [7, 29]. In the finger electrode, MgO$_x$/Al, TiO$_2$/Ca/Al, or TiO$_2$/LiF/Al is widely used as the electron-selective contact. In the rear electrode, MoO$_x$/Ag is widely used as the hole-selective contact. After the optimization processes, the power conversion efficiency (PCE) of PERL c-Si solar cells can be higher than 25%. However, the double-sided fabrication process increases the cost of c-Si solar cells.

To simplify the fabrication process, interdigitated back contact (IBC) c-Si solar cells [30] were proposed to replace PERL c-Si solar cells. Figure 9.2 displays a cross-sectional schematic view of an IBC c-Si solar cell. The surfaces of n-type c-Si substrates can be passivated by Al$_2$O$_3$ thin films added using the ALD method. Compared with p-type c-Si substrates, the chemical properties of n-type c-Si substrates are more stable. Therefore, n-type c-Si substrates are used in IBC solar cells. At the p contact, a Ni/Al electrode is used to collect the photogenerated holes. In between the p+ type Si and the Ni/Al electrode, a SiO$_x$/V$_2$O$_x$ interlayer can be

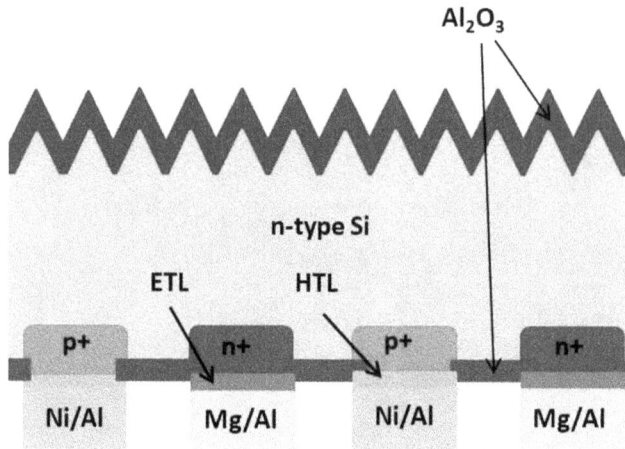

Figure 9.2. Cross-sectional schematic view of an IBC c-Si solar cell.

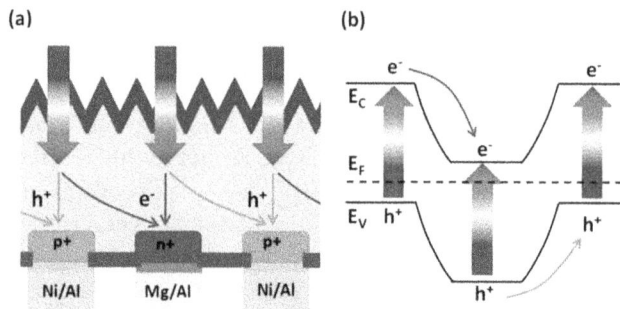

Figure 9.3. The flow paths of photogenerated electrons and holes in an IBC c-Si solar cell.

formed via the thermal evaporation of V_2O_5, which is used as the hole transport layer (HTL). The formation of ultrathin SiO_x is due to the chemical reaction that takes place during the thermal evaporation of V_2O_5. The V_2O_5 can be replaced by other transition-metal oxides (TMOs), such as WO_3 and MoO_3. At the n contact, a Mg/Al electrode is used to collect the photogenerated electrons. In between the n+ type Si and the Mg/Al electrode, an Al_2O_3/TiO_2 interlayer can be formed via the ALD method [31], which is used as the electron transport layer (ETL). The deposition of monolayer Al_2O_3 can be viewed as doping the TiO_2, thereby locally increasing the carrier concentration [32]. The TiO_2 can be replaced by other transparent conductive oxides (TCOs), such as ZnO and SnO_2. Atomic Al can also be used as a dopant to increase the carrier concentration in ZnO and SnO_2. The flow paths of electrons and holes photogenerated by light absorption in an n-type c-Si substrate are plotted in figure 9.3(a), which shows that the photogenerated electrons and holes are spatially separated and thus effectively collected by the Ni/Al cathode and the Mg/Al anode. In the space-energy diagram of n-type Si under sunlight illumination (see figure 9.3(b)), the photogenerated electrons near the p+ region can flow into the adjacent n+ region owing to the sunlight-assisted built-in electric field. For a similar reason, the photogenerated holes near the n+ region can flow into the adjacent p+ region. Compared with PERL c-Si solar cells, the PCE of IBC c-Si solar cells can be higher owing to better carrier collection efficiency and lower carrier recombination. The PCE of IBC c-Si solar cells can be higher than 26%.

HJ Si solar cells are the third kind of c-Si solar cells. Hydrogenated amorphous Si (a-Si:H) thin films were proposed to effectively passivate the top and bottom surfaces of the n-type Si substrate [33]. Figure 9.4 displays a cross-sectional schematic view of an HJ c-Si solar cell. The bandgap values of n-type Si, a-Si:H, and ITO are about 1.12, 1.75, and 4.00 eV, respectively. Figure 9.5 plots the flow paths of electroncs and holes in the space-dependent energy diagram of a HJ c-Si solar cell. After light absorption in the n-type Si substrate, the photogenerated electrons can efficiently tunnel through the narrow barrier of the a-Si:H thin film owing to the high electron density in the n+ type a-Si:H, which is used as the ETL. The photogenerated electrons are then efficiently collected by the ITO in the bottom region of the solar cell owing to the formation of an ohmic contact at the n+ type a-Si:H/ITO interface. In the bottom region, the thickness of the ITO film influences the distribution of light absorption in the n-type Si substrate, which can be designed using the transfer matrix method described in chapter 2. On the rear side, the use of Ag can reduce the series resistance of the resultant solar cell and increase the light-harvesting efficiency of the n-type Si substrate. In addition, the photogenerated holes can efficiently tunnel through the narrow barrier of the a-Si:H thin film owing to the high hole density in the p+ type a-Si:H. The photogenerated holes in the p+ type a-Si:H thin film can then be efficiently collected by the ITO in the top region of the solar cell owing to the formation of an ohmic contact at the p+ type a-Si:H/ITO interface. An Ag finger electrode is used to reduce the series resistance of the resultant solar cell. It should be noted that ITO films are metallic TCOs owing to their high conductivity, which can be greater than 7000 S cm^{-1}. Moreover, the

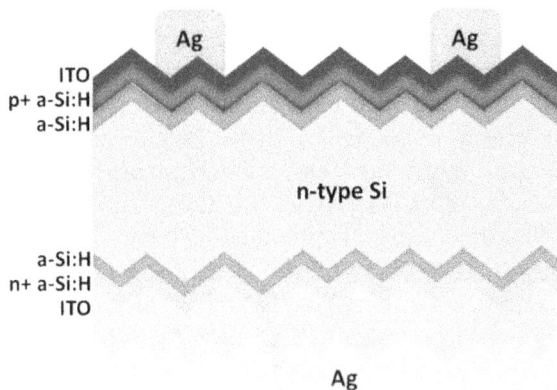

Figure 9.4. Cross-sectional schematic view of an HJ c-Si solar cell.

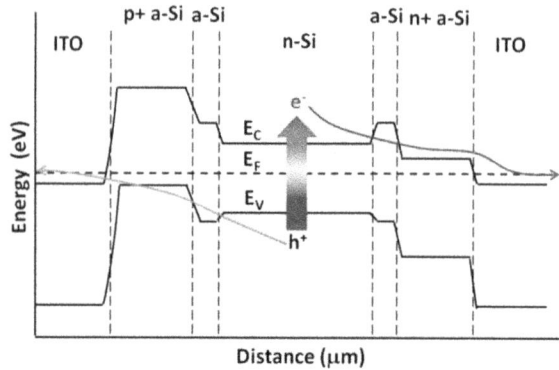

Figure 9.5. The flow paths of electrons and holes in the space-dependent energy diagram of an HJ c-Si solar cell.

optimal thickness of a-Si:H thin films ranges from 5 to 7 nm, which explains the efficient tunneling behaviors of the photogenerated electrons and holes [34]. The highest PCE of HJ c-Si solar cells is close to 27%, mainly due to the superior passivation effects of the a-Si:H thin films on the top and bottom surfaces of the n-type c-Si substrates.

The highest PCEs of PERL, IBC, and HJ c-Si solar cells are about 25%, 26%, and 27%, respectively, which are lower than the theoretical prediction of 33.4% made by the Shockley–Queisser (SQ) limit. When Auger recombination is considered, the PCE of c-Si solar cells is limited to 29.4%. It is, however, possible to reduce the negative influence of Auger recombination on photovoltaic performance by increasing radiative recombination in c-Si, which leads to the prediction that the highest PCE of c-Si solar cells is about 31.6% [35]. In other words, there is room for improvement in the PCE of c-Si solar cells from the current record to the predicted highest value (figure 9.5).

9.3 GaAs solar cells

Gallium arsenide (GaAs) is a direct, narrow-bandgap semiconductor material. Its zinc blende crystal structure, with a 0.565 32 nm lattice constant and a total atomic number of 64, can be used to compute the density of GaAs, which is 5.3176 g cm^{-3}. The direct bandgap energy of GaAs is 1.424 eV, which corresponds to a bandgap wavelength of 871 nm. When the absorption bandgap energy is 1.424 eV, the maximum current density of GaAs solar cells is about 32 mA cm^{-2}. At $\lambda = 800$ nm, the absorption coefficient of GaAs is about 11 885 cm^{-1}. According to the Beer–Lambert law, GaAs can absorb about 90% of the incident photons at $\lambda = 800$ nm when its thickness is 2 μm, which explains the typical thickness of the light-absorbing layer of GaAs solar cells. The highest PCE of GaAs thin-film solar cells is 29.1%, which is close to the highest theoretical value of 32.5% predicted by the SQ limit. The open-circuit voltage (V_{OC}), short-circuit current density (J_{SC}), and fill factor (FF) of an optimal GaAs thin-film solar cell are 1.1272 V, 29.78 mA cm^{-2}, and 86.7%, respectively [36]. Experimental results show that it is possible to increase the

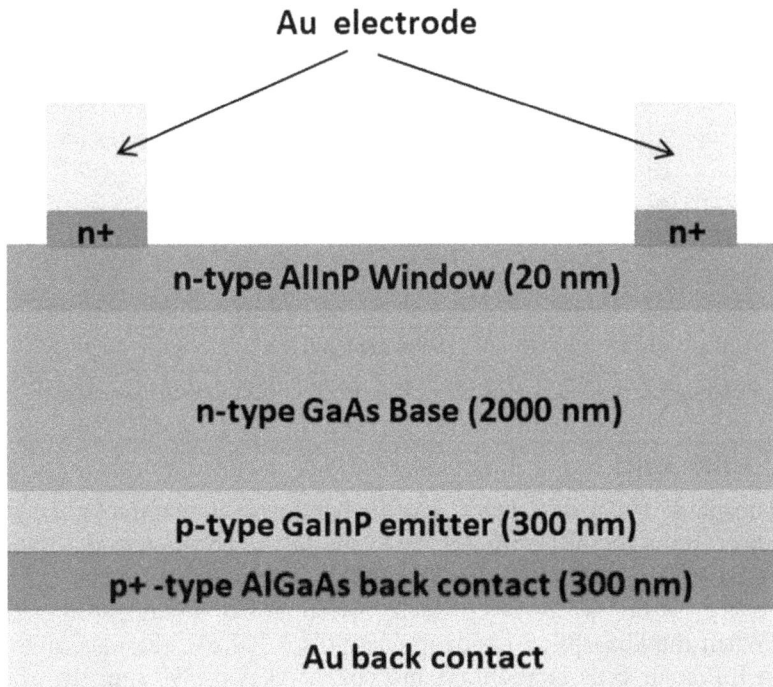

Figure 9.6. Cross-sectional schematic view of an efficient GaAs solar cell without an AR coating layer.

PCE to the SQ limit of GaAs solar cells by enhancing the light-harvesting ability of the GaAs thin film. The method used to calculate the PCE is described in section 9.9.

Figure 9.6 displays a cross-sectional schematic view of an efficient GaAs solar cell without an AR coating layer. An n-type GaAs base is used to absorb the incident sunlight. After light absorption, the photogenerated electrons can be collected by the n-type AlInP window. The 20 nm thick n+ type AlInP forms an ohmic contact between the Au electrode and the n-type AlInP window, thereby efficiently collecting the photogenerated electrons. In addition, the photogenerated holes can be collected by the p-type GaInP emitter. The p+ type AlGaAs forms an ohmic contact with the Au back contact. Figure 9.7 shows that the photogenerated electrons (red arrows) and photogenerated holes (green arrows) can flow to the n+ type AlInP window and the p+ type AlGaAs back contact, respectively, which is mainly due to the stepped energy level structure. At the n+ type AlInP/Au (p+ type AlGaAs/Au) interface, electrons (holes) can efficiently tunnel through the narrow barrier due to the formation of an ohmic contact.

It should be noted that the total thickness of the GaAs solar cell is less than 5000 nm due to the peeled film technology [37]. In other words, the AlInP/GaAs/GaInP/AlGaAs multilayers are fabricated on top of a thick GaAs substrate to form high-quality thin films. After the peeling process, the Au back contact can be formed using the electroplating method [38].

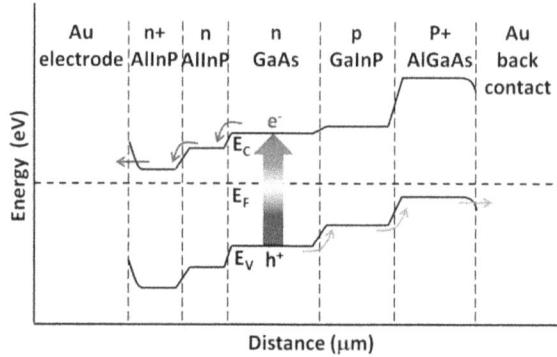

Figure 9.7. Space-dependent energy diagram of an efficient GaAs solar cell.

9.4 InP solar cells

Indium phosphide (InP) is a direct, narrow-bandgap semiconductor material. Its zinc blende crystal structure, with a 0.586 87 nm lattice constant and a total atomic number of 64, can be used to compute its density, which is 4.81 g cm^{-3}. The direct bandgap energy of InP is 1.344 eV, which corresponds to a bandgap wavelength of 923 nm. When the absorption bandgap energy is 1.334 eV, the maximum current density in InP solar cells is about 35 mA cm^{-2}. At $\lambda = 890$ nm, the absorption coefficient of InP is about 15 196 cm^{-1}. According to the Beer–Lambert law, InP can absorb about 90% of the incident photons at $\lambda = 890$ nm when its thickness is 1.5 μm, which explains the typical thickness of the light-absorbing layer of InP solar cells. The highest PCE of InP thin-film solar cells is 24.2%, which is significantly lower than the highest theoretical value of 33.0% predicted by the SQ limit. The V_{OC}, J_{SC}, and FF values of optimal InP thin-film solar cells are 1.008 V, 31.15 mA cm^{-2}, and 82.6%, respectively [39]. Experimental results show that the relatively lower PCE of InP solar cells is mainly due to the larger potential loss, which is proportional to the Urbach energy of the light-absorbing material [40]. Let us compare the material properties of InP and GaAs in order to understand the photovoltaic performance of the resultant solar cells. The Urbach energy values of InP and GaAs are about 7.0 and 5.2 meV at $T = 300$ K, respectively [41]. The temperature-dependent Urbach energy can be written as equation (9.1a), where T is the temperature, $E_U(0)$ is the static component of the Urbach energy, and θ_E is the Einstein phonon temperature [42]. At room temperature, θ_E dominates the Urbach energy. In other words, smaller values of θ_E result in larger values of E_U at room temperature. The Einstein phonon energy (E_{photon}) is proportional to θ_E and the angular frequency (ω), as shown in equation (9.1b), where K_B is the Boltzmann constant and $\hbar$ is the reduced Planck constant. In the spring–mass model, ω can be computed using equation (9.1c), where K is the spring constant and m_{eff} is the effective mass. The total atomic numbers of GaAs and InP are both 64, which means that they have similar effective masses. The lattice constant values of GaAs and InP are 0.565 32 and 0.586 87 nm, respectively, which shows that the spring constant of GaAs is larger than that of InP. In other words, the angular frequency (Einstein phonon energy) of GaAs is higher than that of InP, which explains why the Urbach energy of GaAs is lower than that of InP at $T = 300$ K.

$$E_U(T) = E_U(0) + \frac{2E_U(0)}{e^{\theta_E/T} - 1} \tag{9.1a}$$

$$E_{\text{photon}} = k_B \theta_E = \hbar\omega \tag{9.1b}$$

$$\omega = \sqrt{K/m_{\text{eff}}} \tag{9.1c}$$

9.5 CdTe solar cells

Cadmium telluride (CdTe) is a direct, narrow-bandgap semiconductor material. Its zinc blende crystal structure, with a 0.648 nm lattice constant and a total atomic number of 100, can be used to compute its density, which is 5.85 g cm^{-3}. The direct bandgap energy of CdTe is 1.45 eV, which corresponds to a bandgap wavelength of 855 nm. When the absorption bandgap energy is 1.45 eV, the maximum current density of CdTe solar cells is about 31 mA cm^{-2}. At $\lambda = 800$ nm, the absorption coefficient of CdTe is about 13 085 cm^{-1}. According to the Beer–Lambert law, CdTe can absorb about 90% of the incident photons at $\lambda = 800$ nm when its thickness is 1.75 µm, which explains the typical thickness of the light-absorbing layer of CdTe solar cells. The highest PCE of CdTe thin-film solar cells is 21.0%, which is significantly lower than the highest theoretical value of 32.0% predicted by the SQ limit. The V_{OC}, J_{SC}, and FF values of optimal InP thin-film solar cells are 1.0623 V, 30.25 mA cm^{-2}, and 79.4%, respectively [43]. The Urbach energy of CdTe thin films is about 50 meV [44], which is far larger than the Urbach energy of GaAs thin films. The larger Urbach energy of the CdTe thin films explains the relatively lower V_{OC} values of the resultant solar cells.

9.6 CIGS solar cells

CIGS thin films are composed of Cu, In, Ga, and Se. Their chemical formula is written as $CuIn_xGa_{1-x}Se$, where x can vary from zero to one. The bandgap energy of CIGS thin films decreases from 1.7 to 1.0 eV as the x value increases from zero to one. Conceptually, there is a trade-off between the V_{OC} and J_{SC} values of CIGS solar cells at different x values. When the x value is larger, V_{OC} and J_{SC} are higher and lower, respectively. In solar cells, the Urbach energy and bandgap energy of the widely used CIGS thin films are about 20 meV and 1.5 eV, respectively [45]. Therefore, the PCE of CIGS thin-film solar cells is predicted to be lower than that of CdTe thin-film solar cells. It has been found that a high concentration of Ag can be used in CIGS thin films to achieve a high PCE of 23.64% [46]; the resultant solar cells are called ACIGS thin-film solar cells. The V_{OC}, J_{SC}, and FF values of optimal ACIGS thin-film solar cells are 0.767 V, 38.3 mA cm^{-2}, and 80.5%, respectively. It should be noted that the bandgap energy and the average grain diameter of multicrystalline ACIGS thin films are about 1.130 eV and 2000 nm, respectively.

The grain diameter is close to the thickness of the ACIGS thin film, which means that the photogenerated electrons and holes can be effectively collected by an n-type CdS ETL and a p-type $MoSe_2$ HTL, respectively. The Urbach energy of ACIGS thin films is about 14.5 meV at $T = 300$ K [46], which is slightly higher than the Urbach energy of c-Si. In multicrystalline thin films, the Urbach energy can be decreased by increasing the grain size [47, 48]. Therefore, it is predicted that the photovoltaic performance of ACIGS thin-film solar cells can be improved by decreasing the Urbach energy of ACIGS thin films.

9.7 CZTSSe solar cells

CZTSSe thin films are composed of Cu, Zn, Sn, S, and Se. Their chemical formula is written as $Cu_2ZnSn(S_xSe_{1-x})_4$, where x can vary from zero to one. The bandgap energy of CZTSSe thin films decreases from 1.5 to 1.0 eV as the x value decreases from one to zero. Conceptually, there is a trade-off between the V_{OC} and J_{SC} values of CZTSSe solar cells at different x values. When the x value is larger, V_{OC} and J_{SC} are higher and lower, respectively. The highest PCE of CZTSSe thin-film solar cells is 13.14% when the bandgap energy and Urbach energy of the CZTSSe thin films are 1.11 eV and 22.7 meV, respectively [49]. In the best CZTSSe thin-film solar cell, the multilayered device architecture is glass/Mo/GeO_2/CZTSSe/CdS/ZnO/ITO with a Ni/Al finger electrode; GeO_2 and CdS are used as the HTL and ETL, respectively. At the Mo/GeO_2 and ZnO/ITO interfaces, ohmic contacts can be formed. The V_{OC}, J_{SC}, and FF values of the best CZTSSe solar cells are 0.5472 V, 34.3 mA, and 70%, respectively. It should be noted that GeO_2 is used to form a Ge-Se liquid flux that facilitates nucleation and grain growth during the fabrication of the CZTSSe thin film, thereby resulting in a flatter thin film with larger grains. The Urbach energy (bandgap energy) values of the CZTSSe thin films and c-Si are about 22.7 meV (1.11 eV) and 13.0 meV (1.12 eV) at $T = 300$ K, respectively. Therefore, the Urbach energy of the CZTSSe thin films is significantly larger than that of c-Si at $T = 300$ K, which explains why the V_{OC} of the best CZTSSe thin-film solar cells is significantly lower than that of efficient c-Si solar cells. This means that it is possible to effectively increase the PCE of CZTSSe thin-film solar cells by decreasing the Urbach energy of CZTSSe thin films.

9.8 MAPbI$_3$ solar cells

Methylammonium lead triiodide ($MAPbI_3$) is a direct, narrow-bandgap organic–inorganic semiconductor material. The lattice constants a, b, and c of tetragonal $MAPbI_3$ are 0.886, 0.886, and 1.266 nm [50], respectively, which correspond to bandgap energies of about 1.63 eV. The PL emission peak wavelength of $MAPbI_3$ thin films is about 770 nm. When the absorption bandgap energy is 1.63 eV, the maximum current density of $MAPbI_3$ thin-film solar cells is about 25 mA cm^{-2}. The absorption coefficient of $MAPbI_3$ thin films is highly related to their crystallinity. Therefore, it is difficult to determine the ideal thickness of multicrystalline $MAPbI_3$ thin films used in solar cells solely by examining the absorption coefficient at wavelengths near the absorption edge. According to previous reports, the J_{SC} values can be higher than 22 mA cm^{-2} when the thickness values of MAPbI$_3$ thin films range from 300 to 500 nm [51–53]. The highest PCE values of regular-type and inverted-type MAPbI$_3$ thin-film solar cells are both higher than 20%.

It should be noted that high-quality MAPbI$_3$ thin films can be prepared using solution processes at low thermal annealing temperatures (less than 140 °C). When the spin-coating method is used to prepare MAPbI$_3$ thin films, white crystals of CH$_3$NH$_3$I (MAI) and brown powdered PbI$_2$ are dissolved in a highly polar solvent mixture to form the MAPbI$_3$ precursor solution. Widely used solvents are gamma-butyrolactone (GBL), dimethyl sulfoxide (DMSO), dimethylformamide (DMF), and N-methyl-2-pyrrolidone (NMP). During the spin-coating process, a low-polarity solvent is added to the MAPbI$_3$ precursor solution as an antisolvent; this is called the washing-enhanced nucleation (WEN) process [54]. This key method is also called a fast deposition-crystallization procedure [55] or a solvent engineering method [56]. Widely used antisolvents include chlorobenzene, toluene, and diethyl ether. The fabrication conditions of the spin-coating method and the WEN process can be used to manipulate the grain size of MAPbI$_3$ thin films from about 200 to 2000 nm, significantly influencing the optoelectronic properties of the resultant thin films.

In regular-type MAPbI$_3$ solar cells, the device architecture is glass/fluorine-doped tin oxide (FTO)/ETL/MAPbI$_3$/HTL/Au. Thin films made from TiO$_2$ and SnO$_2$ nanoparticles are two widely used ETLs. When a thin film of SnO$_2$ is used as the ETL, the FTO can be replaced by ITO due to the lower required fabrication temperatures, which are below 200 °C. When a thin film of TiO$_2$ is used as the ETL, FTO must be used owing to the higher sintering temperatures required for the formation of anatase TiO$_2$ crystals, which range from 400 °C to 500 °C. ZnO, Al-doped ZnO (AZO), and C$_{60}$ thin films can also be used as the ETL of regular-type MAPbI$_3$ solar cells. Molecular thin films of 2,2',7,7'-tetrakis(N,N-di-p-methoxy-phenyl amino)-9,9'-spirobifluorene (Spiro-OMeTAD) are widely used as the HTL of regular-type MAPbI$_3$ solar cells. The hole mobility of the Spiro-OMeTAD thin films must be increased by adding lithium salts in order to effectively collect photo-generated holes from the MAPbI$_3$ thin films. Au thin films can form ohmic contacts with Spiro-OMeTAD thin films. Au thin films can also be used as efficient reflectors in the wavelength range from 650 to 760 nm, thereby increasing the light-harvesting ability of MAPbI$_3$ thin films. In an efficient regular-type MAPbI$_3$ solar cell, the device architecture is glass/FTO/c-TiO$_2$/mp-TiO$_2$/MAPbI$_3$/Spiro-OMeTAD/Au, where c-TiO$_2$ is compact TiO$_2$ and mp-TiO$_2$ is mesoporous TiO$_2$. The PCE of the regular-type MAPbI$_3$ thin-film solar cells can be greatly increased from 18.51% to 22.12% due to the formation of a strain-relaxed tetragonal MAPbI$_3$ thin film when the surface of mp-TiO$_2$ is coated with a thin layer of CsPbBr$_3$ [57]. The V_{OC}, J_{SC}, and FF values of the efficient regular-type MAPbI$_3$ are 1.12 V, 24.2 mA cm^{-2}, and 81.5%, respectively. Figure 9.8 displays an energy diagram of a regular-type MAPbI$_3$ thin-film solar cell. The bandgap energy of the MAPbI$_3$ thin film is about 1.6 eV. After light absorption takes place in the MAPbI$_3$ thin film, the Fermi levels for electrons and holes (E_F^n and E_F^p) are close to the conduction band minimum (E_{CBM}) and the valence band maximum (E_{VBM}), respectively. In other words, E_F^n and E_F^p are close to E_{CBM} and E_{VBM}, respectively, thereby minimizing the reduction

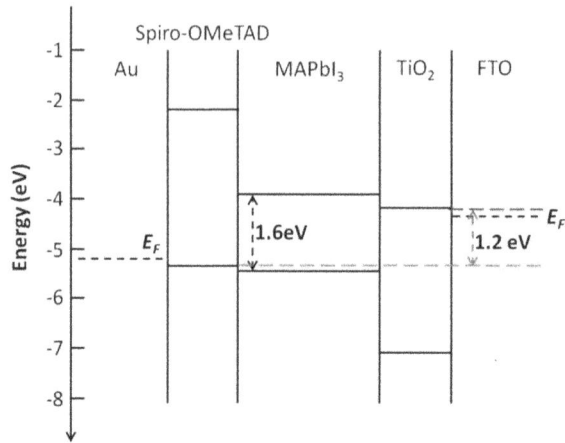

Figure 9.8. Space-dependent energy diagram of a regular-type MAPbI$_3$ thin-film solar cell.

in V_{OC}. The energy difference between the E_{CBM} value of TiO$_2$ and the lowest unoccupied molecular orbital (LUMO) of Spiro-OMeTAD is about 1.2 eV, which is close to the V_{OC} value of the optimal regular-type MAPbI$_3$ solar cell. In other words, ohmic contacts form at the Spiro-OMeTAD/Au and TiO$_2$/FTO interfaces, thereby minimizing voltage loss. It is still possible to improve V_{OC} by increasing the energy difference between the E_{CBM} value of the ETL and the E_{VBM} value of the HTL. However, there is a trade-off between V_{OC} and J_{SC} that limits the PCE of the resultant solar cells.

In inverted-type MAPbI$_3$ solar cells, the device architecture is glass/ITO/HTL/MAPbI$_3$/ETL/Ag. Poly(triaryl amine) (PTAA), (2-(3,6-dimethoxy-9H-carbazol-9-yl)ethyl)phosphonic acid (MeO-2PACz), and poly[3-(4-carboxybutyl)thiophene-2,5-diyl] (P3CT) are widely used hole transport materials in the HTL. PTAA and P3CT are p-type polymers. The LUMO values of PTAA and P3CT thin films are about −5.25 eV, which is slightly higher than the E_{VBM} value of MAPbI$_3$ thin films. In addition, the LUMO value of MeO-2PACz small molecule thin films is about −5.3 eV. Therefore, the photogenerated holes in MAPbI$_3$ thin films can be collected by p-type organic thin layers while slightly reducing the potential energy of the holes. C$_{60}$ and PCBM are two widely used ETL materials in inverted-type MAPbI$_3$ solar cells. The LUMO values of C$_{60}$ and PCBM thin films are both about −3.90 eV, which is slightly lower than the E_{CBM} value of MAPbI$_3$ thin films. Conceptually, the photogenerated electrons in MAPbI$_3$ thin films can be collected by n-type organic thin films. However, these can form an electrical barrier between the hydrophilic MAPbI$_3$ thin film and the hydrophobic C$_{60}$ (PCBM) thin film, which greatly decreases the electron collection efficiency. Fortunately, the quality of the MAPbI$_3$/PCBM (MAPbI$_3$/C$_{60}$) interface and the PCBM (C$_{60}$) thin film can be improved by a bathocuproine (BCP)/isopropanol (IPA) treatment process [58]. In the BCP/IPA treatment process, the BCP and IPA molecules both penetrate into the PCBM (C$_{60}$) thin film and the PCBM/MAPbI$_3$ (C$_{60}$/MAPbI$_3$) interface. BCP molecules that penetrate the PCBM (C$_{60}$) thin films can passivate the disordered molecules, thereby reducing the electron trap density. At the PCBM/MAPbI$_3$ (C$_{60}$/ MAPbI$_3$) interface, BCP molecules can passivate electron-poor defects, thereby reducing the formation of electrical barriers. It should be noted that the use of the BCP/IPA treatment can greatly increase the PCE of MAPbI$_3$ solar cells from about 15% to 20%. When an efficient inverted-type MAPbI$_3$ solar cell was measured in the reverse scan direction, its V_{OC}, J_{SC}, and FF values were 1.105 V, 22.43 mA cm^{-2}, and 79.4%, respectively [58]. The HTL and ETL used were a thin layer of P3CT-Na and a thin film of BCP:PCBM, respectively. When the device was measured in the forward scan direction, the FF significantly increased from 79.4% to 82.6% while maintaining the V_{OC} and J_{SC} values, resulting in an improvement in the PCE from 19.68% to 20.49%. This means that shallow defects in MAPbI$_3$ thin films can be partially filled in the reverse scan direction, thereby greatly increasing the FF. It should be noted that the relatively lower J_{SC} value can be explained by the formation of ripples in the external quantum efficiency (EQE) spectrum. In the EQE spectrum of inverted-type MAPbI$_3$ thin-film solar cells, the formation of ripples is due to thin-film interference between the glass/ITO and ITO/HTL interfaces. In other words, antireflective measures should be considered to reduce reflections from the inter-faces, thereby increasing the J_{SC} value of the resultant inverted-type MAPbI$_3$ solar cells while maintaining high EQE values in the absorption wavelength range from 300 to 800 nm.

MAPbI$_3$ is an outstanding photoactive material that can efficiently convert sunlight to electricity owing to its superior optoelectronic properties and its large tolerable deviation in its crystal structure. However, its lack of chemical stability limits its practical applications. MAPbI$_3$ can be degraded via the dehydrogenation process, which forms CH$_3$NH$_2$, HI, and PbI$_2$ when PbI$_6$ octahedra participate in chemical reactions. The chemical equation for the degradation can be written as equation (9.2a). At room temperature, CH$_3$NH$_2$, HI, and PbI$_2$ are gaseous, liquid, and solid, respectively. Additionally, C$_{60}$ or the small molecules of PCBM can penetrate the MAPbI$_3$ thin film along the grain boundaries, resulting in the formation of CH$_3$NH$_2$-C$_{60}$-CH$_3$NH$_2$ cations in between adjacent MAPbI$_3$ grains, thereby forming large-bandgap HPbI$_3$ perovskites in the upper regions of MAPbI$_3$ thin films [53, 59, 60]. The chemical equation can be written as equation (9.2b). The two chemical reactions can occur when the materials are heated and/or illuminated. When C$_{60}$ (PCBM) molecules are located near the grain boundaries, the chemical reaction only produces solids, which increases the stability of the inverted-type MAPbI$_3$ thin-film solar cells. However, the formation of large-bandgap HPbI$_3$ perovskites results in an electrical barrier between the MAPbI$_3$ thin film and the C$_{60}$ (PCBM) thin film, creating s-shaped characteristics in the J–V curves of the encapsulated solar cells, thereby increasing their series resistance [5, 53].

$$CH_3NH_3PbI_3 \rightarrow CH_3NH_2 + HI + PbI_2 \tag{9.2a}$$

$$2CH_3NH_3PbI_3 + C_{60} \rightarrow CH_3NH_2(C_{60})CH_3NH_2 + 2HPbI_3 \tag{9.2b}$$

According to theoretical predictions, the highest PCE of MAPbI$_3$ solar cells is about 30.5%, which motivates us to improve the chemical stability of MAPbI$_3$ thin-film solar cells. The two chemical equations above show that it is possible to eliminate the chemical reaction of the $MAPbI_3$ thin film in an encapsulated solar cell by avoiding the dehydrogenation process. C_{60} molecules can passivate the electron-rich defects of $MAPbI_3$ thin films due to the electron-accepting ability of the conjugated carbon ring structure. However, interfacial MA cations are electron-poor defects in $MAPbI_3$ thin films, which can react with the C_{60} molecules via the pressure-activated and light-driven dehydrogenation process. In other words, it is possible to eliminate the chemical reaction shown in equation (9.2b) when the C_{60} molecules are replaced by non-fullerene n-type molecules as the ETL of inverted-type $MAPbI_3$ thin-film solar cells. Furthermore, other research has shown that the chemical stability of organic–inorganic perovskite thin films can be greatly improved when the $MAPbI_3$ perovskite is replaced by formamidinium lead iodide ($FAPbI_3$) perovskite. However, the high density of cracks in $FAPbI_3$ thin films limits their practical applications in thin-film-based optoelectronic devices. These cracks are due to the cubic crystal structure, which leads to larger thermal expansion coefficients than those of the substrates used [61].

9.9 Evaluating solar cells

The photovoltaic performance of solar cells can be determined by measuring and analyzing the current density–voltage (J–V) curve under one sun of illumination. In general, the intensity and spectrum of the sunlight used are 100 mW cm^{-2} and AM 1.5G, respectively. During the measurement, the solar cell is illuminated at normal incidence. Figure 9.9(a) displays the measurement circuit connected to a solar cell, where R is a small resistor that is usually less than $0.1\ \Omega$ in a commercially sourced meter system. During the measurement, a voltage is applied to the solar cell at the input port, and the current is measured at the output port with a small resistor R. When the applied voltage is zero, the measured photocurrent (I_{ph}) is defined as a negative value, as shown in figure 9.9(b). When the active area (A) of the solar cell is known, the absolute value of the photocurrent density (I_{ph}/A) is denoted by J_{SC}. Photogenerated electrons and holes appear at the cathode and anode of the solar cell, respectively. The generation of photocurrent can be understood by examining the space-dependent energy diagram of a solar cell. In figure 9.8, the electrons and

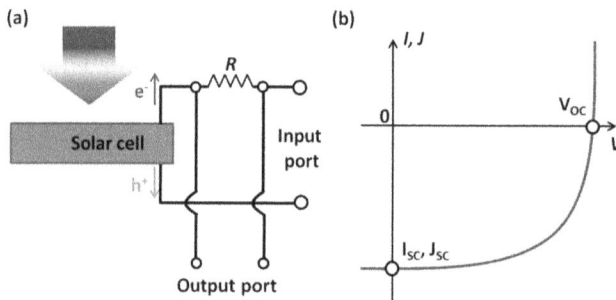

Figure 9.9. (a) Measurement circuit used to evaluate solar cells. (b) Typical J–V curve of a solar cell under one sun of illumination.

holes photogenerated in the light-absorbing layer (MAPbI$_3$) are collected by the ETL/cathode and HTL/anode, respectively, due to the driving forces from the built-in electric fields and the formation of ohmic contacts. When a voltage is applied to the solar cell, the applied voltage decreases the collection efficiency of the photo-generated electrons and holes. When the photocurrent value is zero, the applied voltage is defined as the V_{OC} of the measured solar cell. To evaluate the perfection of the solar cell, the FF is computed by using equation (9.3a), where P_{max} is the maximum power density value of the J–V curve (not shown). A higher shunt resistance at the J_{SC} point and a lower series resistance at the V_{OC} point result in a larger FF. The PCE of a solar cell can then be computed using equation (9.3b), where I_{sun} is the sunlight intensity (100 mW cm^{-2}, AM 1.5).

$$FF = P_{max}/(V_{OC} \times J_{SC}) \qquad (9.3a)$$

$$PCE = \frac{V_{OC} \times J_{SC} \times FF}{I_{sun}} \qquad (9.3b)$$

The hysteresis characteristics in the J–V curve are related to defects in the solar cells. Figure 9.10 displays the V_{OC} hysteresis and the J_{SC} hysteresis. The V_{OC} hysteresis is mainly related to defects at the HTL/light-absorbing layer interface and/or the light-absorbing layer/ETL interface. In other words, the presence of V_{OC} hysteresis means that the contact quality is imperfect at the interfaces. The J_{SC} hysteresis is mainly related to bulk defects in the light-absorbing layer. In polycrystalline lead trihalide perovskite thin films, iodide vacancies inside the crystals are the main bulk defects. Conceptually, larger values of J_{SC} hysteresis in the J–V curve indicate that the light-absorbing layer has a broader PL peak width and larger Urbach energy. In other words, the presence of bulk defects in the light-absorbing layer can be confirmed by analyzing the J–V curve, PL spectrum, and absorbance spectrum.

The light intensity-dependent V_{OC} curves of solar cells can be used to evaluate the trend of defect densities using equation (9.4a), where J_S is the saturation current density under reverse bias, $K_B T$ is the thermal energy, q is the charge of an electron, and $S(k_B T/q)$ is defined as the slope of the $\ln(J_{SC})$–V_{OC} curve [62]. It should be

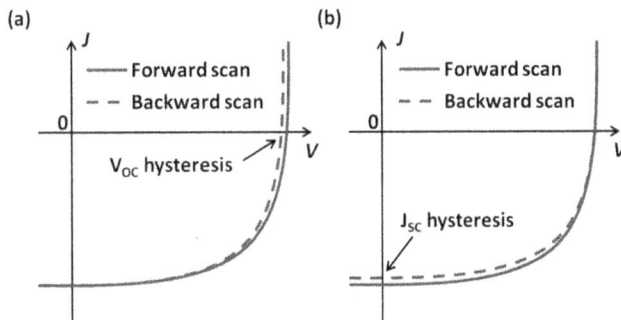

Figure 9.10. Hysteresis in solar cells: (a) V_{OC} hysteresis and (b) J_{SC} hysteresis.

noted that the small J_S value can be ignored in equation (9.4a). To simplify the relation between V_{OC} and J_{SC}, the equation can be written as equation (9.4b), which is called the Koster relation. Before using the slope as an indicator to qualitatively evaluate the defect density, the linear relation between light intensity and J_{SC} must be satisfied. In an imperfect p–n junction-based solar cell, the slope value is larger than one. A larger slope value corresponds to a higher defect density in the solar cell. Therefore, the slope of the Koster relation can be used to evaluate the defect density in the measured solar cell.

$$V_{OC} = S\frac{k_B T}{q} \operatorname{Ln}\left(\frac{J_{SC}}{J_S + 1}\right) \tag{9.4a}$$

$$V_{OC} = S(k_B T/q)\operatorname{Ln}(J_{SC}) \tag{9.4b}$$

Bibliography

[1] Schinke C et al 2015 Uncertainty analysis for the coefficient of band-to-band absorption of crystalline silicon *AIP Adv.* **5** 67168

[2] Green M A 2013 Improved value for the silicon free exciton binding energy *AIP Adv.* **3** 112104

[3] Aspnes D E, Kelso S M, Logan R A and Bhat R 1986 Optical properties of $Al_xGa_{1-x}As$ *J. Appl. Phys.* **60** 754–67

[4] Nam S B, Reynolds D C, Litton C W, Almassy R J and Collins T C 1976 Free-exciton energy spectrum in GaAs *Phys. Rev.* B **13** 761–7

[5] Aspnes D E and Studna A A 1983 Dielectric functions and optical parameters of Si, Ge, GaP, GaAs, GaSb, InP, InAs, InSb from 1.5 to 6.0 eV *Phys. Rev.* B **27** 985–1009

[6] Turner W J, Reese W E and Pettit G D 1964 Exciton absorption and emission in InP *Phys. Rev.* **136** A1467

[7] Treharne R E, Seymour-Pierce A, Durose K, Hutchings K, Roncallo S and Lane D 2011 Optical design and fabrication of fully sputtered CdTe/CdS solar cells *J. Phys. Conf. Ser.* **286** 012038

[8] Strzalkowski K, Zakrzewski J and Malinski M 2013 Determination of the exciton binding energy using photothermal and photoluminescence spectroscopy *Int. J. Thermophys.* **34** 691–700

[9] Fluegel B, Alberi K, DiNezza M J, Liu S, Zhang Y-H and Mascarenhas A 2014 Carrier decay and diffusion dynamics in single-crystalline CdTe as seen via microphotoluminescence *Phys. Rev. Appl.* **2** 034010

[10] Sharbati S, Gharibshahian I and Orouji A A 2019 Designing of $Al_xGa_{1-x}As$/CIGS tandem solar cell by analytical model *Sol. Energy* **188** 1–9

[11] Gokmen T, Gunawan O and Mitzi D B 2013 Minority carrier diffusion length extraction in $Cu_2ZnSn(Se,S)_4$ solar cells *J. Appl. Phys.* **114** 114511

[12] Mohammadnejad S and Parashkouh A B 2017 CZTSSe solar cell efficiency improvement using a new band-gap grading model in absorber layer *Appl. Phys.* A **123** 758

[13] Ball J M et al 2015 Optical properties and limiting photocurrent of thin-film perovskite solar cells *Energy Environ. Sci.* **8** 602–9

[14] Niedzwiedzki D M, Zhou H and Biswas P 2022 Exciton binding energy of MAPbI$_3$ thin film elucidated via analysis and modeling of perovskite absorption and photoluminescence properties using various methodologies *J. Phys. Chem.* C **126** 1046–54

[15] Scajev P, Miasojedovas S and Jursenas S 2020 A carrier density dependent diffusion coefficient, recombination rate and diffusion length in MAPbI$_3$ and MAPbBr$_3$ crystals measured under one- and two-photon excitations *J. Mater. Chem.* C **8** 10290–301

[16] Stelling C, Singh C R, Karg M, Konig T A F, Thelakkat M and Retsch M 2017 Plasmonic nanomeshes: their ambicalent role as transparent electrodes in organic solar cells *Sci. Rep.* **7** 42530

[17] Xiong C, Sun J, Cai C, Caiyang W and Zhu Y 2020 Disclosing exciton binding energy of organic materials from absorption spectrum *Sol. Energy* **204** 155–60

[18] Chang S H, Chiang C-H, Chen H-M, Tai C-Y and Wu C-G 2013 Broadband charge transfer dynamics in P3HT:PCBM blended film *Opt. Lett.* **38** 5342–5

[19] Arefinia Z and Samajdar D P 2021 Novel semi-analytical optoelectronic modeling based on homogenization theory for realistic plasmonic polymer solar cells *Sci. Rep.* **11** 3261

[20] Kroh D, Athanasopoulos S, Nadazdy V, Kahle F-J, Bassler H and Kohler A 2023 An impedance study of the density of states distribution in blends of PM6:Y6 in relation to barrierless dissociation of CT states *Adv. Funct. Mater.* 2302520

[21] Tokmoldin N, Hosseini S M, Raoufi M, Phuong L Q, Sandberg O J, Guan H, Zhou Y, Neher D and Shoaee S 2020 Extraordinarily long diffusion length in PM6:Y6 organic solar cells *J. Mater. Chem.* A **8** 7854–60

[22] Zhao J, Wang A, Altermatt P P, Wenham S R and Green M A 1996 24% Efficient perl silicon solar cell: recent improvements in high efficiency silicon cell research *Sol. Energy Mater. Sol. Cells* **41** 87–9

[23] Manea E, Budianu E, Purica M, Cristea D, Cernica I, Muller R and Poladian V M 2005 Optimization of front surface texturing processes for high-efficiency silicon solar cells *Solar Energy Mater. Solar Cells* **87** 423–31

[24] Chebotareva A B, Untila G G and Kost T N 2015 Heavily phosphorus-doped silicon nanoparticles as intermediate layer in solar cell based on IFO/p-Si heterojunction *Sol. Energy* **122** 650–7

[25] Sheng J, Ma Z, Cai W, Ma Z, Ding J, Yuan N and Zhang C 2019 Impact of phosphorus diffusion on n-type poly-Si based passivated contact silicon solar cells *Sol. Energy Mater. Sol. Cells* **203** 110120

[26] Fair R B 1975 Boron diffusion in silicon-concentration and orientation dependence, background effects, and profile estimation *J. Electrochem. Soc.: Solid-State Sci. Technol.* **122** 801–5

[27] Lelievre J-F, Kafle B, Saint-Cast P, Brunet P, Magnan R, Hernandez E, Pouliquen S and Massines F 2019 Efficient silicon nitride SiN$_x$:H antireflective and passivation layers deposited by atmospheric pressure PECVD for silicon solar cells *Prog. Photovolt.* **27** 1007–9

[28] Theeuwes R J, Melskens J, Black L E, Beyer W, Koushik D, Berghuis W J H, Macco B and Kessels W M M 2021 PO$_x$/Al$_2$O$_3$ stacks for c-Si surface passivation: material and interface properties *ACS Appl. Electron. Mater.* **3** 4337–47

[29] Richter A, Patel H, Reichel C, Benick J and Glunz S W 2023 Improved silicon surface passivation by ALD Al$_2$O$_3$/SiO$_2$ multilayers with *in situ* plasma treatments *Adv. Mater. Interfaces* **10** 2202469

[30] Masmitja G, Ortega P, Puigdollers J, Gerling L G, Martin I, Voz C and Alcubilla R 2018 Interdigitated back-contacted crystalline silicon solar cells with low-temperature dopant-free selective contacts *J. Mater. Chem.* A **6** 3977–85

[31] Szindler M, Szindler M M, Orwat J and Kulesza-Matlak G 2022 The Al_2O_3/TiO_2 double antireflection coating deposited by ALD method *Opto-Electron. Rev.* **30** e141952

[32] Said N D M, Sahdan M Z, Ahmad A, Senain I, Bakri A S, Abdullah S A and Rahim M S 2017 Effects of Al doping on structural, morphology, electrical and optical properties of TiO_2 thin film *AIP Conf. Proc.* **1788** 030130

[33] Ge J, Ling Z P, Wong J, Mueller T and Aberle A G 2012 Optimsation of intrinsic a-Si:H passivation layers in crystalline-amorphous silicon heterojunction solar cells *Energy Procedia* **15** 107–17

[34] Balaji P, Dauksher W J, Bowden S G and Augusto A 2020 Improving surface passivation on very thin substrates for high efficiency silicon heterojunction solar cells *Solar Energy Mater. Solar Cells* **216** 110715

[35] Ehrler B, Alarcon-Llado E, Tabernig S W, Veeken T, Garnett E C and Polman A 2020 Photovoltaics reaching for the Schockley–Queisser limit *ACS Energy Lett.* **5** 3029–33

[36] Green M A, Dunlop E D, Hohl-Ebinger J, Yoshita M, Kopidakis N and Hao X 2020 Solar cell efficiency tables (version 56) *Prog. Photovolt.* **28** 629

[37] Konagai M, Sugimoto M and Takahashi K 1978 High efficiency GaAs thin film solar cells by peeled film technology *J. Cryst. Growth* **45** 277–80

[38] Tang H, Chen C-Y, Nagoshi T, Chang T-F M, Yamane D, Machida K, Masu K and Sone M 2016 Enhancedment of mechanical strength in Au films electroplated with supercritical carbon dioxide *Electrochem. Commun.* **72** 126–30

[39] Green M A, Hishikawa Y, Warta W, Dunlop E D, Levi D H, Hohl-Ebinger J and Ho-Haillie A W H 2017 Solar cell efficiency tables (version 52) *Prog. Photovolt.* **25** 668–76

[40] Chantana J, Kawano Y, Nishimura T, Mavlonov A and Minemoto T 2020 Impact of Urbach energy on open-circuit voltage deficit of thin-film solar cells *Solar Energy Mater. Solar Cells* **210** 110502

[41] Ledinsky M, Schonfeldova T, Holovsky J, Aydin E, Hajkova Z, Landova L, Neykova N, Fejfar A and De Wolf S 2019 Temperature depencence of the Urbach energy in lead iodide perovskites *J. Phys. Chem. Lett.* **10** 1368–73

[42] Grein C H and John S 1989 Temperature dependence of the Urbach optical absorption edge: A theory of multiple phonon absorption and emission sidebands *Phys. Rev.* B **39** 1140

[43] Green M A, Dunlop E D, Yoshita M, Mopidakis N, Bothe K, Siefer G and Hao X 2023 Solar cell efficiency tables (version 62) *Prog. Photovolt.* **31** 651–63

[44] Amin N, Karim M R and Alothman Z A 2021 Impact of $CdCl_2$ treatment in CdTe thin film grown on ultra-thin glass substrate via close spaced sublimation *Crystals* **11** 390

[45] Chantana J, Kawano Y, Nishimura T and Minemoto T 2019 Urbach energy of Cu(In,Ga) Se_2 and Cu(In,Ga)(S,Se)$_2$ absorbers prepared by various methods: indicator of their quality *Mater. Today Commun.* **21** 100652

[46] Keller J, Kiselman K, Donzel-Gargand O, Martin N M, Babucci M, Lundberg O, Wallin E, Stolt L and Edoff M 2024 High-concentration silver alloying and steep back-contact gallium grading enabling copper indium gallium selenide solar cell with 23.6% efficiency *Nat. Energy* **9** 467–78

[47] Chiang S-E, Wu J-R, Cheng H-M, Hsu C-L, Shen J-L, Yuan C-T and Chang S H 2020 Origins of the s-shape characteristics in J–V curve of inverted-type perovskite solar cells *Nanotechnoloyg* **31** 115403

[48] Thakur D *et al* 2021 Structural, optical and excitonic properties of urea grading doped $CH_3NH_3PbI_3$ thin films and their application in inverted-type perovskite solar cells *J. Alloys Compd.* **858** 157660

[49] Wang J *et al* 2022 Ge bidirectional diffusion to simultaneously engineer back interface and bulk defects in the absorber for efficient CZTSSe solar cells *Adv. Mater.* **34** 2202858

[50] Guo L, Xu G, Tang G, Fang D and Hong J 2020 Structural stability and optoelectronic properties of tetragonal $MAPbI_3$ under strain *Nanotechnology* **31** 225204

[51] Ying B *et al* 2015 Perovskite solar cells with near 100% internal quantum efficiency based on large single crystalline grains and vertical bulk heterojunctions *J. Am. Chem. Soc.* **137** 9210–3

[52] Ke Q B, Wu J-R, Lin C-C and Chang S H 2022 Understanding the PEDOT:PSS, PTAA, P3CT-X hole-transport-layer-based inverted perovskite solar cells *Polymers* **14** 823

[53] Chandel A, Ke Q B, Chiang S-E, Cheng H-M and Chang S H 2023 Effects of dyring time on the formation of merged and soft $MAPbI_3$ grains and their photovoltaic responses *Nanoscale Adv.* **5** 2190–8

[54] Chang S H, Wong S-D, Huang H-Y, Yuan C-T, Wu J-R, Chiang S-E, Tseng Z-L and Chen S-H 2019 Effects of the washing-enhanced nucleation process on the material properties and performance of perovskite solar cells *J. Alloys Compd.* **808** 151723

[55] Xiao M, Huang F, Huang W, Dkhissi Y, Zhu Y, Etheridge J, Gray-Weale A, Bach U, Cheng Y-B and Spiccia L 2014 A fast deposition-crystallization procedure for highly efficient lead iodide perovskite thin-film solar cells *Angew. Chem.* **53** 9898–903

[56] Jeon N J, Noh J H, Kim Y C, Yang W S, Ryu S and Seok S I 2014 Solvent engineering for high-performance inorganic-organic hybrid perovskite solar cells *Nat. Mater.* **13** 897–903

[57] Ye T, Wang K, Ma S, Wu C, Hou Y, Yang D, Wang K and Priya S 2021 Strain-relaxed tetragonal $MAPbI_3$ results in efficient mesoporous solar cells *Nano Energy* **83** 105788

[58] Chiang S-E *et al* 2021 On the role of solution-processed bathocuproine in high-efficiency inverted perovskite solar cells *Sol. Energy* **218** 142–9

[59] Thakur D, Chiang S-E, Yang M-H, Wang J-S and Chang S H 2022 Self-stability of un-encapsulated polycrystalline MAPbI3 solar cells via the formation of chemical bonds between C60 molecules and MA cations *Solar Energy Mater. Solar Cells* **235** 111454

[60] Thakur D, Ke Q B, Chiang S-E, Tseng T-H, Cai K-B, Yuan C-T, Wang J-S and Chang S H 2022 Stable and efficient soft perovskite crystalline film based solar cells prepared with a facile encapsulation method *Nanoscale* **14** 17625

[61] Huang Y-C, Yen I-J, Tseng C-H, Wang H-Y, Chandel A and Chang S H 2024 Structural and excitonic properties of the polycrystalline $FAPbI_3$ thin films, and their photovoltaic response *Nanotechnology* **35** 505706

[62] Koster L J A, Mihailetchi V D, Ramaker R and Blom P W 2005 Light intensity dependence of open-circuit voltage of polymer:fullerene solar cells *Appl. Phys. Lett.* **86** 123509

IOP Publishing

Light–Material Interactions and Applications in Optoelectronic Devices

Anjali Chandel and Sheng Hsiung Chang

Chapter 10

Light-emitting diodes

The first part of this chapter graphically describes the main light emission mechanisms in semiconductors, which are photoluminescence, phosphorescence, thermally activated delayed fluorescence (TADF), and exciplex emissions. There are two time constants in TADF emission, which are assigned to singlet exciton and triplet exciton lifetimes, respectively. It should be noted that an exciplex can be formed at a donor/acceptor interface owing to the accumulated electrons and holes. The energy of exciplexes can be transferred to the energy states of photoluminescence or phosphorescence, which are defined as exciplex hosts. In the second part, the main working mechanisms of InGaN, $CsPbBr_3$ perovskite quantum dots, TADF molecules, near-infrared (NIR) phosphors, and exciplex light-emitting diodes (LEDs) are described and discussed, which can be used to understand how to achieve the highest external quantum efficiency at different emission wavelengths.

10.1 Light emission mechanisms in semiconductors

After excitation, the radiative relaxations of electrons from the excited state to the ground state can be classified into fluorescence, phosphorescence, and exciplex emission, which follow the Pauli exclusion principle. In other words, the instantaneous spin state of the electrons influences the light emission efficiency. When the abovementioned three emissions occur in a light-emitting diode, the wavelength of exciplex emission (phosphorescence) is longer than that of phosphorescence (fluorescence), which can be used to form a white light source or a broadband emitter. Figure 10.1 displays the energy diagram of a multiemission system. When the electrons in the S_1 state of the acceptor and the holes in the highest occupied molecular orbital (HOMO) of the donor are radiatively recombined, the emission wavelength is the shortest among the three emissions, which is called exciplex emission at a donor/acceptor interface. In the acceptor molecule, electrons in the S_1

doi:10.1088/978-0-7503-6099-9ch10

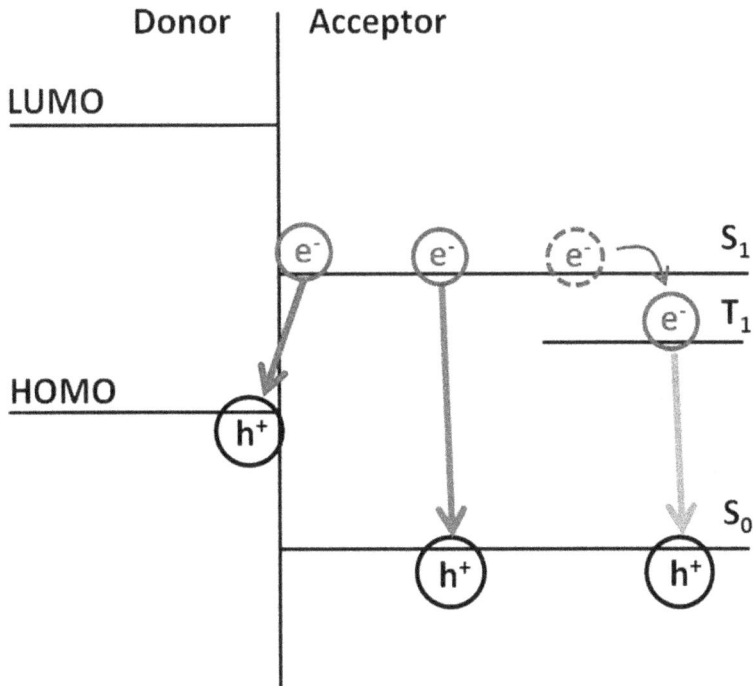

Figure 10.1. Energy diagram of a multiemission system.

state can radiatively recombine with holes in the S_0 state, which is called edge emission in emissive semiconductor materials. In organic semiconductor materials, electrons can efficiently relax to the T_1 state from the S_1 state via the intersystem crossing (ISC) process, thereby forming non-emissive and emissive triplet excitons.

In inorganic semiconductor materials, light emission is mainly produced by band-to-band transitions and defect-assisted transitions. When electrons are trapped in defects in inorganic semiconductor materials, they have two relaxation pathways: radiative decay and non-radiative decay. Therefore, the formation of non-emissive defect states must be minimized in order to increase the light emission efficiency of inorganic semiconductor materials. The exciton binding energy values of inorganic semiconductor materials are less than 50 meV; therefore, excitons can form free carriers, resulting in low light emission efficiency. After light absorption (exciton generation) in a semiconductor material, the generation efficiency of free carriers can be evaluated using equation (6.4a). The transition from an exciton (electron–hole pair) to a free electron and a free hole is called dissociation. After dissociation, the distance between the electron and hole is larger than the exciton radius, which indicates that the attractive force between free electrons and free holes is weaker than that between the electron and hole of an exciton. When carrier diffusion occurs, the light emission efficiency of radiative recombination between free electrons and holes is low owing to the weak electron–hole connection. In other words, exciton dissociation can be minimized by spatially confining the electrons and holes.

In inorganic LEDs, quantum wells (QWs) and quantum dots (QDs) have been used to increase carrier confinement, which can greatly increase the electroluminescence intensity of the resultant devices. The theories of QWs and QDs are described in chapter 8.

In lead trihalide perovskites, the emission wavelength is strongly related to the halide radius: a larger halide radius corresponds to a lower energy bandgap. $CsPbBr_3$ and γ-phase $CsPbI_3$ perovskites have been used to realize highly efficient LEDs. The exciton binding energy values of $CsPbBr_3$ and γ-phase $CsPbI_3$ perovskites are about 38 and 33 meV at room temperature, respectively [1, 2]. Compared to β-phase $MAPbI_3$ perovskites, the γ-phase $CsPbI_3$ perovskites have a higher exciton binding energy, which indicates that Cs lead trihalide perovskite QDs have great potential for use as light-emissive materials in LEDs. To form perovskite QDs, benzylammonium (BA^+) and phenethylammonium (PEA^+) cations have been used to terminate the perovskite grains. Figure 10.2 displays the molecular structures of BAI and PEAI. The iodide anions of BAI and PEAI can passivate the electron-poor defects of the perovskite grains. BA^+ and PEA^+ can occupy cation positions at the surfaces of the perovskite grains during the crystal growth process, thereby determining the physical sizes of the perovskite QDs. It is noted that the exciton binding energy is largely influenced by the surface-terminated cations when the physical size of the perovskite QDs is reduced to the thickness of quasi-2D perovskites, which is mainly due to the different dielectric screening strengths of the surface-terminated cations [3].

In organic semiconductor materials, the emitted light waves are mainly related to singlet-state exciton emission and triplet-state exciton emission. In general, singlet-state excitons can emit light waves which correspond to the energy difference between the lowest unoccupied molecular orbital (LUMO) and the HOMO. In the triplet state (T_1), electrons cannot radiatively decay to the ground state (S_0) due to the spin-forbidden transition. Fortunately, the electrons in the triplet state (T_1) can radiatively decay to the ground state (S_0) after spin flipping, owing to the spin–orbit coupling in metal-based complexes [4]. In addition, the fluorescent intensity of organic semiconductor materials can be increased using the reverse intersystem crossing (RISC) process. In other words, electrons in the triplet state (T_1) can transfer to the singlet state (S_1), thereby increasing the fluorescent intensity of organic semiconductor materials. Figure 10.3 displays electron transfers using energy diagrams of organic materials and metal-based complexes. In organic

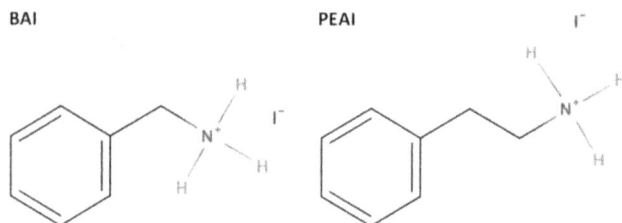

Figure 10.2. Molecular structures of BAI and PEAI.

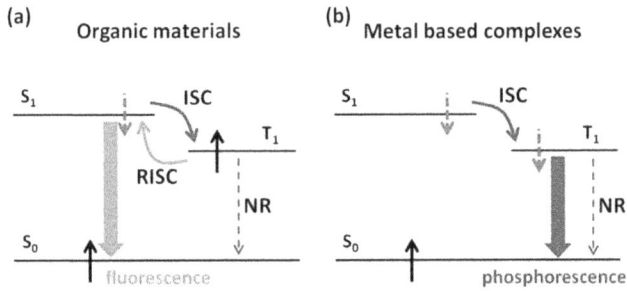

Figure 10.3. Energy diagrams of organic materials and metal-based complexes. (a) Fluorescence; (b) phosphorescence.

semiconductor materials, RISC cannot occur when the energy difference between the S_1 state and the T_1 state is too large, resulting in non-emissive triplet excitons and therefore leading to non-radiative decay. When the energy difference between the S_1 state and the T_1 state is less than 100 meV, the efficient RISC process can greatly increase the fluorescent intensity, which is called TADF [5]. It should be noted that ISC and RISC in organic semiconductor materials flip the spin state of the electrons from spin down (up) to spin up (down). Time-resolved photoluminescence (TRPL) can be used to confirm the occurrence of the TADF process in organic semi-conductor materials. In organic materials that support TADF, the TRPL curves have two time constants of about 10 ns and 1 μs. The faster and slower relaxation times correspond to singlet exciton and triplet exciton lifetimes, respectively. In metal-based complexes, the ISC process can maintain the spin state of the electrons from the S_1 state to the T_1 state, which results in efficient phosphorescence. Iridium (Ir), europium (Eu), platinum (Pt), and copper (Cu) complexes have been widely used as efficient phosphorescent materials [6–9].

10.2 InGaN-based LEDs

The development of efficient and high-power blue inorganic LEDs has played an important role in display technologies and underwater communications. The near-band-edge (NBE) emission wavelength of GaN is about 370 nm. Figure 10.4(a) displays the energy diagram of a homojunction p–i–n GaN structure. The electrons and holes are injected from the n-GaN and p-GaN into the i-GaN, respectively, which forms accumulated electron–hole pairs in the i-GaN layer, thereby emitting light waves from the NBE transition. The red dashed circle and green dashed circle denote accumulated electrons and holes, respectively, which indicates a weak overlap between the electron and hole (e–h) wave functions, thereby limiting the light emission efficiency. Figure 10.4(b) displays the energy diagram of a hetero-junction p-AlGaN/InGaN/n-GaN structure, which can result in a better overlap between the e–h wave functions in the InGaN layer. However, there is no hole confinement ability at the InGaN/p-AlGaN interface, which limits the light emission efficiency. In other words, the light emission efficiency can be improved by increasing the overlap of e–h wave functions with an appropriate design that

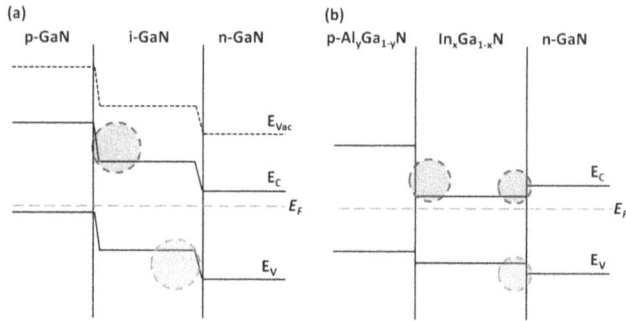

Figure 10.4. Energy diagrams. (a) A homojunction p–i–n GaN structure. (b) A heterojunction p-AlGaN/i-InGaN/n-GaN structure.

includes energy-level engineering while maintaining the carrier injection efficiency [10, 11].

In figure 10.4(b), the $In_xGa_{1-x}N$ layer has smaller energy bandgap values and performs carrier localization. The energy bandgap of $In_xGa_{1-x}N$ decreases from 3.44 to 0.62 eV as the x value increases from zero to one, which can be used to vary the emission wavelength of InGaN-based LEDs over a wide range from about 370 nm to about 2000 nm. The lattice constants and thermal expansion coefficients of wurtzite AlN, GaN, and InN are listed in table 8.1. The lattice mismatch values at the GaN/AlN and InN/GaN interfaces are about -2.50% and -9.89%, respectively. The small lattice mismatch value at the GaN/AlN interface explains why AlGaN alloy compounds are used as a buffer layer to reduce the residual stress in the GaN layer, thereby decreasing the dislocation density. However, the large lattice mismatch value at the InN/GaN interface explains the relatively lower emission efficiency of red InGaN LEDs [12]. It should be noted that the in-plane thermal expansion coefficients of GaN and InN are $5.59 \times 10^{-6} \text{ K}^{-1}$ and $3.80 \times 10^{-6} \text{ K}^{-1}$, respectively, which means that the lattice constants of $In_xGa_{1-x}N$ and GaN can be matched at higher temperatures, thereby reducing the formation of vacancies in the $In_xGa_{1-x}N$ layers of the multiple $In_xGa_{1-x}N$/GaN QWs. In other words, the crystal growth temperature can greatly influence the quality of the multiple $In_xGa_{1-x}N$/GaN QWs.

10.3 Efficient blue InGaN LEDs

Figure 10.5 displays a cross-sectional schematic view of an efficient blue InGaN LED. AlGaN is grown on an Al_2O_3 substrate as the buffer layer. The use of a patterned Al_2O_3 surface and an AlGaN buffer layer can reduce the dislocation density in the subsequent epitaxial layers. n-GaN and p-AlGaN are used to inject electrons and holes into the multiple InGaN/GaN QWs, respectively. Inverted cones are formed by an etching process that removes defects near dislocations. An SiO_2 layer is used to passivate the surface defects. The formation of cones and the use of the SiO_2 layer effectively reduce the ideality factor of the resultant LEDs [13]. In the QW region, the metal forms an ohmic contact with the p-type AlGaN. In the

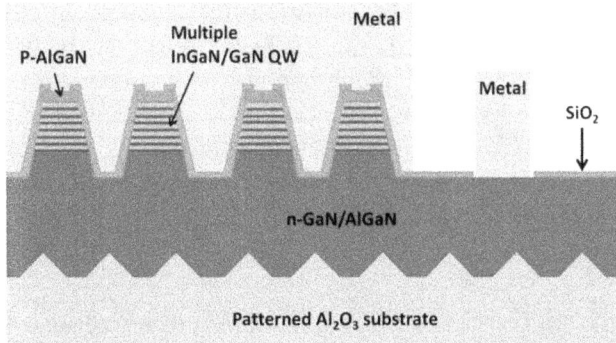

Figure 10.5. Cross-sectional schematic view of an efficient blue InGaN LED.

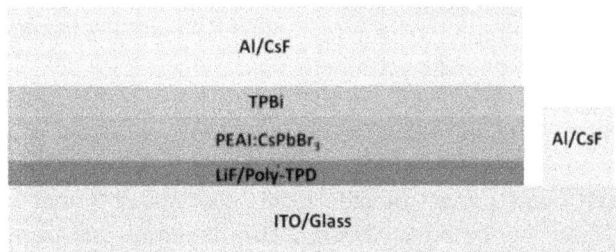

Figure 10.6. Cross-sectional schematic view of an efficient PEAI:CsPbBr3 QD LED.

non-QW region, the metal forms an ohmic contact with the n-type GaN. Blue electroluminescence is emitted from the bottom surface of the Al_2O_3 substrate. At a wavelength of 450 nm, the refractive index of Al_2O_3 is about 1.7794, which results in a critical angle of 34.2° at the Al_2O_3/air interface, thereby limiting the light extraction efficiency owing to total internal reflection. The light extraction efficiency can be greatly improved by using a two-dimensional array of dielectric spheres [14].

10.4 Efficient $CsPbBr_3$ QD LEDs

Figure 10.6 displays a cross-sectional schematic view of an efficient PEAI:$CsPbBr_3$ QD LED [15]. A 20 nm thick layer of poly[N,N′-bis(4-butylphenyl)-N,N′-bis (phenyl)-benzidine] (poly-TPD) and a 40 nm thick layer of 1,3,5-tri(1-phenyl-1H-benzimidazol-2-yl)benzene (TPBi) are used as the hole injection layer (HIL) and the electron injection layer (EIL), respectively. A 3 nm thick LiF layer is used to increase the hydrophilicity of the substrate, which can enhance the uniformity of the PEAI: $CsPbBr_3$ QD thin film. A 1.5 nm thick CsF layer is used as a buffer layer to increase the electron injection efficiency at the Al/CsF interface. In the light-emissive layer, the anions and cations of PEAI molecules can occupy the vacancies of the electron-poor and electron-rich defects of the $CsPbBr_3$ QDs, respectively. However, the surface defects of $CsPbBr_3$ QDs cannot be completely passivated by the anions and cations of the PEAI molecules. When 18-crown-6 molecules and poly(ethylene glycol) methyl ether acrylate (MPEG-MAA) polymers are added to the PEAI:

10-6

CsPbBr$_3$ QD thin film, the carrier lifetime in the QDs can be significantly increased from about 19 ns to about 90 ns due to the better defect passivation ability of the additives used. In addition, the Urbach energy value is decreased from about 40 meV to about 23 meV when the two additives are used, which means that the crystallinity of the PEAI:CsPbBr$_3$ QDs can be improved by passivating the surface defects. It should be noted that the maximum external quantum efficiency (EQE) of the optimized PEAI:CsPbI$_3$ QD LED is 28.1%, which occurs at a wavelength of 513 nm.

10.5 Thermally activated delayed fluorescence organic LEDs

The device architecture of TADF organic light-emitting diodes (OLEDs) is similar to that of the efficient CsPbBr$_3$ QD LEDs (see figure 10.6) because the two devices can be fabricated using the same methods, which mainly include spin-coating and vacuum thermal evaporation. In general, the light emission efficiency of TADF OLEDs is mainly related to the RISC efficiency, since RISC converts the energy from the non-emissive triplet excitons into emissive singlet excitons. Many metal-containing organic molecules exhibit efficient TADF based on a combination of electron-deficient and electron-rich atoms. Figure 10.7(a) displays the molecular structure of a well-known green molecule, namely 2,6-bis(9H-carbazol-9-yl)boron (CzBN). The carbon-ring-based molecule CzBN contains one boron (B) and two nitrogen (N) atoms, which are electron deficient and electron rich, respectively [16, 17]. In other words, the LUMO and HOMO have a large spatial overlap, thereby resulting in a larger exciton binding energy in CzBN-based molecules. Figure 10.7(b) displays the molecular structure of a modified CzBN molecule with four electron-donating groups (electron donors) and one electron-withdrawing group (electron acceptor). The use of these electron donors and electron acceptor can greatly increase the HOMO energy level and decrease the LUMO energy level, which results in a decrease in the bandgap energy of CzBN-based molecules. The emission peak wavelength of CzBN-based molecules is red-shifted from about 490 nm to about 640 nm when the molecules are modified with the electron-donating and electron-withdrawing groups simultaneously [18].

The device structure of a highly efficient CzBN-based TADF LED is Al/Liq/ ETL-2:Liq/DMFBD-TRZ:NPB:CzBN/PEDOT:PSS/ITO/glass, where Al forms the anode, 8-hydroxyquinolinolato-lithium (Liq) is the EIL, ETL:Liq is the electron

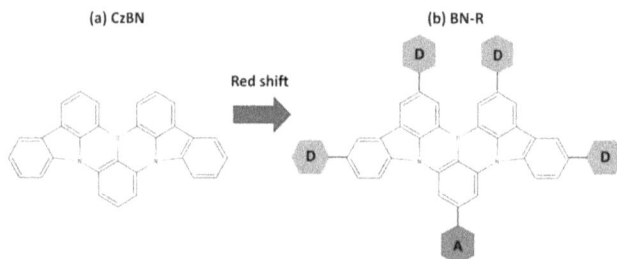

Figure 10.7. Molecular structure of TADF molecules. (a) CzBN; (b) BN-R.

transport layer, 2-(3'-(9,9-dimethyl-9H-fluoren-2-yl)-[1,1'-bi-phenyl]-3-yl)-4,6-diphenyl-1,3,5-triazine (DMFBD-TRZ) and N4,N4'-di(naphthalene-1-yl)-N4,N4'-diphenyl-[1,1'-biphenyl]-4,4'-diamine (NPB) are host materials, the CzBN-based molecule is the TADF emitter, poly(3,4-ethylenedioxythiophene):polystyrene sulfonate (PEDOT:PSS) is the HIL, and indium tin oxide (ITO) is the transparent anode electrode. It should be noted that the PEDOT:PSS is modified with a perfluorinated ionomer (PFI), which effectively increases the carrier concentration in the PEDOT chains, thereby increasing the hole injection efficiency. The highest EQE of the CzBN-based TADF LEDs is about 22% at a wavelength of 617 nm.

10.6 Efficient near-infrared phosphor LEDs

Ions of rare-earth elements, such as Er^{3+}, Er^{2+}, Eu^{3+}, Eu^{2+}, Nd^{3+}, Nd^{2+}, Yb^{3+}, and Yb^{2+}, can efficiently emit NIR light waves owing to the extremely long carrier lifetime of their radiative recombination. These rare-earth elements can be doped into SiO_2, GeO_2, Al_2O_3, ZnO, and CaO, thereby replacing the cations of the oxides, which results in ionic rare earths as the color centers. The carrier lifetime of phosphorescence can be greatly decreased by aggregation due to quenching when the dopant concentrations of the rare-earth elements are too high, which indicates that the energy transfer between the color centers increases the excited carrier–phonon interaction rate and thereby decreases the light emission efficiency. The photogenerated carrier lifetime also decreases as the doping concentration of rare-earth elements increases in the phosphorescent oxides. Therefore, the concentrations of rare-earth elements as the dopants of phosphorescent oxides have upper limits.

Figure 10.8 displays a cross-sectional schematic view of an efficient NIR phosphor LED that emits light waves at a central wavelength of 760 nm under blue light excitation due to the use of Eu^{2+} ions in CaO as color centers [19]. CaO:Eu

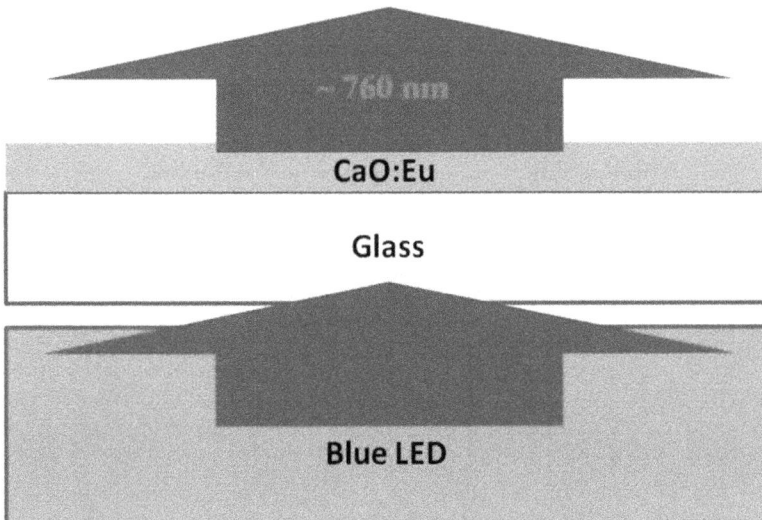

Figure 10.8. Cross-sectional schematic view of an efficient NIR phosphor LED.

powders are mixed with epoxy resin, which is deposited on top of a glass substrate to form a uniform film. It should be noted that the width of the NIR phosphorescence spectrum is about 120 nm, which indicates that CaO:Eu can be used as a gain medium in a femtosecond laser system with an ultrashort pulse duration. In general, a pulse width of 10 nm corresponds to a pulse duration of 100 fs, which follows the time–bandwidth product relation: $\Delta\lambda \times \Delta T = C$, where C is a constant related to the pulse shape [20]. The C values of a Gaussian pulse and a hyperbolic secant pulse are about 0.441 and 0.315, respectively, which can be computed using Fourier transform calculations. In other words, a pulse width of 120 nm corresponds to a pulse duration of about 8 fs in a complete dispersion compensation system.

10.7 Electroplex LEDs

Exciplex emissions can be formed at the donor/acceptor interface, which results in emission at a longer wavelength. In a single-molecular-emitter white organic LED (WOLED), the device architecture in the vertical direction can be Al/Cs$_2$CO$_3$/TmPyPB/Ir(dmppy)$_2$(dpp)/TmPyPB/TAPC/HAT-CN/ITO/glass, where Ir(dmppy)$_2$(dpp) is the designed emitter of the WOLED [21]. The molecular structures of Ir(dmppy)2(dpp), 1,1-Bis[(di-4-tolylamino)phenyl]cyclohexane (TAPC), and 1,3,5-Tris(3-pyridyl-3-phenyl)benzene (TmPyPB) are plotted in figure 10.9(a). The electrons and holes are injected by an Al electrode and an ITO electrode, respectively. Cs$_2$CO$_3$ and 1,4,5,8,9,11-hexaazatriphenylene hexacarbonitrile (HAT-CN) are used to decrease the injection barriers for electrons and holes, respectively. The Cs$_2$CO$_3$ and HAT-CN form the EIL and HIL of the LED, respectively. At the TmPyPB/TAPC interface, the accumulated electrons and holes form electroplexes, which can emit light waves at longer wavelengths. The molecular structures of TmPyPB and TAPC show that the

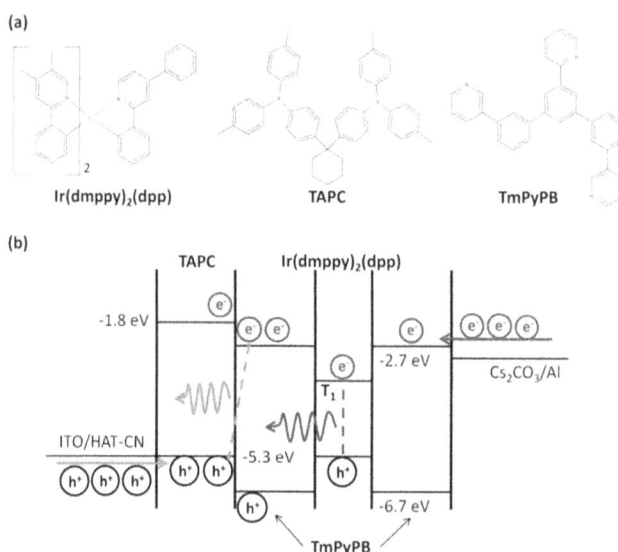

Figure 10.9. (a) Molecular structures. (b) Energy diagram of a single-molecule-emitter white organic LED.

two small molecules can have close contact owing to π–π stacking between aromatic rings. Figure 10.9(b) displays a simplified energy diagram of the single-molecular-emitter WOLED. After carrier injection takes place, electrons and holes can be localized in the Ir(dmppy)2(dpp) molecules, thereby resulting in yellow emission from the T_1 triplet state. In addition, the injected electrons and holes can be localized at the TmPyPB/TAPC interface, which forms electroplexes owing to potential barriers for the electrons and holes. The emission spectra show that different TmPyPB/TAPC interfaces can be formed, thereby resulting in blue and red emissions. Therefore, the blue, yellow, and red emissions can form a white light source. It should be noted that the bandgap values of the TAPC and TmPyPB molecules are larger than 3 eV, which indicates that the visible light emission is not due to radiative recombination in these large-bandgap molecules.

Bibliography

[1] Yuan Y, Chen M, Yang S, Shen X, Liu Y and Cao D 2020 Exciton recombination mechanism in solution grown single crystalline $CsPbBr_3$ perovskite *J. Lumin.* **226** 117471

[2] Gau D L, Galain I, Aguiar I and Marotti R E 2023 Origin of photoluminescence and experimental determination of exciton binding energy, exciton–phonon interaction, and urbach energy in γ-$CsPbI_3$ nanoparticles *J. Lumin.* **257** 119765

[3] Filip M R, Qiu D Y, Ben M D and Neaton J B 2022 Screening of excitons by organic cations in quasi-two-dimensional organic-inorganic lead-halide perovskites *Nano Lett.* **22** 4870–8

[4] Shafikov M Z, Zaytsev A V and Kozhevnikov V N 2021 Halide-enhaced spin–orbit coupling and the phosphorescence rate in Ir(III) complexes *Inorg. Chem.* **60** 642–50

[5] Dias F B, Penfold T J and Monkman A P 2017 Photophysics of thermally activated delayed fluorescence molecules *Methods Appl. Fluoresc.* **5** 012001

[6] You Y and Park S Y 2009 Phosphorescent iridium(III) complexes: toward high phosphorescence quantum efficiency through ligand control *Dalton Trans.* **8** 1267–82

[7] Wang D, Luo Z, Liu Z, Wang D, Fan L and Yin G 2016 Synthesis and photoluminescent properties of Eu (III) complexes with fluorinated b-diketone and nitrogen heterocylic ligands *Dyes Pigm.* **132** 398–404

[8] Li K, Tong G S M, Wan Q, Cheng G, Tong W-Y, Ang W-H, Kwong W-L and Che C-M 2016 Highly phosphorescent platinum(II) emitters: photophysics, materials and biological applications *Chem. Sci.* **7** 1653–73

[9] Wen C, Tao G, Xu X, Feng X and Luo R 2011 A phosphorescent copper(I) complex: synthesis, characterization, photophysical property, and oxygen-sensing behavior *Spectrochim. Acta* A **79** 1345–51

[10] Zhang Z-H, Kyaw Z, Liu W, Ji Y, Wang L, Tan S T, Sun X W and Demir H V 2015 A hole modulator for InGaN/GaN light-emitting diodes *Appl. Phys. Lett.* **106** 063501

[11] Sirkeli V P, Yilmazoglu O, Al-Daffaie S, Oprea I, Ong D S, Kuppers F and Hatnagel H L 2017 Efficiency enhancement of InGaN/GaN light-emitting diodes with pin-doped GaN quantum barrier *J. Phys. D: Appl. Phys.* **50** 035108

[12] Ewing J J, Lynsky C, Wong M S, Wu F, Chow Y C, Shapturenka P, Iza M, Nakamura S, Denbaars S P and Speck J S 2023 High external quantum efficiency (6.5%) InGaN V-defect LEDs as 600 nm on patterned sapphire substrates *Opt. Express* **31** 41351–60

[13] Wong M S, Lee C, Myers D J, Hwang D, Kearns J A, Li T, Speck J S, Nakamura S and DenBaars S P 2019 Size-independent peak efficiency of III-nitride micro-light-emitting-diodes using chemical treatment and sidewall passivation *Appl. Phys. Express* **12** 097004

[14] Chen Y-T, Chandel A, Wu J-R and Chang S H 2021 Coupled periodically electric quadrupoles assisted ultra-broadband lightr-extraction enhancement of red GaN LEDs by manipulating the field orthogaonality at nanoscales *Chin. J. Phys.* **7** 188–95

[15] Liu Z *et al* 2021 Perovskite light-emitting diodes with EQE exceeding 28% through a synergetic dual-additive strategy for defect passivation and nanostructure regulation *Adv. Mater.* **33** 2103268

[16] Yin X, Liu J and Jakle F 2021 Electron-deficient conjugated materials via p–p* conjugation with boron: extending monomers to oligomers, macrocycles, and polymers *Chem. Eur. J.* **27** 2973–86

[17] Yan C *et al* 2020 Synthesis and properties of hypervalent electron-rich pentacoordinate nitrogen compounds *Chem. Sci.* **11** 4082–5088

[18] Cai X, Xu Y, Pan Y, Li L, Pu Y, Zhuang X, Li C and Wang Y 2023 Thermally activated delayed fluorescence emitter for organic light-emitting diode with external quantum efficiency over 20% *Angew. Chem. Int. Ed.* **62** e20221647

[19] Qiao J, Zhang S, Zhou X, Chen W, Gautier R and Xia Z 2022 Near-infrared light-emitting diodes utilizing a europium-activated calcium oxide phosphor with external quantum efficiency of up to 54.7% *Adv. Mater.* **34** 2201887

[20] Lazaridis P, Debarge G and Gallion P 1995 Time-bandwitt product of chirped sech^2 pulses: application to phase-amplitude-coupling factor measurement *Opt. Lett.* **20** 1160–2

[21] Luo D, Li X-L, Zhao Y, Gao Y and Liu B 2017 High-performance blue molecular emitter-free and doping-free hybrid white organic light-emitting diodes: an alternative concept to manipulate charges and excitons based on exciplex and electroplex emission *ACS Photonics* **4** 1566–75